Practical Support for ISO 9001
Software Project Documentation

Practical Support for ISO 9001 Software Project Documentation

Using IEEE Software Engineering Standards

Susan K. Land

John W. Walz

IEEE COMPUTER SOCIETY

WILEY-INTERSCIENCE

A WILEY-INTERSCIENCE PUBLICATION

Published by John Wiley & Sons, Inc., Hoboken, New Jersey
Published simultaneously in Canada.

For general information on our other products and services please contact our Customer Care Department within the U.S. at 877-762-2974, outside the U.S. at 317-572-3993 or fax 317-572-4002.

Wiley also publishes its books in a variety of electronic formats. Some content that appears in print, however, may not be available in electronic format.

Library of Congress Cataloging-in-Publication Data is available.

ISBN-13 978-0-471-76867-8
ISBN-10 0-471-76867-7

Printed in the United States of America.

10 9 8 7 6 5 4 3 2 1

Contents

Preface

The IEEE Computer Society Software and Systems Engineering Standards Committee (S2ESC) is the governing body responsible for the development of software and systems engineering standards. S2ESC has conducted several standards users' surveys. The results of these surveys revealed that standards users found the most value in the guides and standards that provided the specific detail that they needed for the development of their process documentation. Users consistently responded that they used the guides in support of software process definition and improvement (ISO 9001 or CMMI®) but that these standards and guides required considerable adaptation when applied as an integrated set of software process documentation.

This book was written to support software engineering practitioners who are responsible for producing the process documentation, and work products or artifacts, associated with support of software process definition and improvement. This book will be most useful to organizations with multiple products and having business customer relationships. In addition to members of project development and test teams working on products with multiple versions, this book is also useful to members of organizations supporting software project development and testing, such as project management, configuration management, risk management, human resources, and information technology.

It is the hope of the authors that this book will help members of organizations who are responsible for developing or maintaining their software processes in order to support ISO 9001 documentation requirements (ISO 9001:2000, Quality Management Systems—Requirements).

Software process definition, documentation, and improvement should be an integral part of every software engineering organization. This book addresses the specific documentation requirements in support of ISO 9001 by providing detailed documentation guidance in the form of:

- Detailed organizational policy examples.
- An integrated set of over 40 deployable document templates.
- Examples of over 100 common work products required in support of assessment activities.
- Examples of organizational delineation of process documentation.

This book provides a set of templates based on IEEE software engineering standards that support the documentation required for all activities associated with software development projects. The goal is to provide practical support for individuals responsible for the development and documentation of software processes and procedures. The objective is to present the reader with an integrated set of documents that support the requirements of ISO 9001.

It is hoped that this book will provide specific support for organizations pursuing software process definition and improvement. For organizations that do not wish to pursue ISO 9001 accreditation, this text will show how the application of IEEE standards can facilitate the development of sound software engineering practices.

ACKNOWLEDGMENTS

Susan K. Land
I would like to acknowledge my company, Northrop Grumman Information Technology TASC, and thank them for their continued support of my IEEE Computer Society volunteer activities. In these days of continued corporate cutbacks, I feel privileged to work within an organization that supports standardization and the pursuit of software engineering excellence. I would also like to acknowledge my colleagues within the volunteer organizations of the IEEE Computer Society and thank them for their constant encouragement, their dedication to quality, and their friendship. I would specifically like to thank James Moore, J. Fernando Naveda, and Alan Clements. I would like to thank my husband for his unwavering support and encouragement and my father for his years of good advice and good example.

John W. Walz
Without the support of Ann, my wife, this book would not have been a reality—thank you. I would like to acknowledge my company, The Sutton Group, and thank Stan Flowers for his continued support of my IEEE Computer Society volunteer activities. Also I would like to thank my previous managers who guided my professional development: Thomas J. Scurlock, Jr. and Terry L. Welsher, both retired from Lucent Technologies, and James R. McDonnell, SBC. For my long involvement in software engineering standards, I would like to thank the leadership of Helen M. Wood.

Both authors would like to thank Andrew Prince of John Wiley & Sons for his outstanding ability see past each grammatical error and into what we really meant to say. We would also like to thank Angela Burgess and Deborah Plummer of the IEEE Computer Society for their support and encouragement.

Chapter 1

Introduction and Overview

INTRODUCTION

Moving an organization from the chaotic environment of free-form development toward a more controlled and documented process can be overwhelming to those tasked to make it happen. This book is written to support software engineering practitioners who are responsible for producing the process documentation and work products or artifacts associated in support of software process definition and improvement. This book specifically addresses how IEEE standards may be used to facilitate the development of processes, internal plans, and procedures in support of managed and defined software and systems engineering processes for conformance to ISO 9001 [47]. This text describes how IEEE Software Engineering Standards can be used to help support the development and definition of corporate best practices.

The IEEE Computer Society Software and Systems Engineering Standards Committee (S2ESC) offers an integrated suite of standards based on using IEEE 12207 [39] as a set of reference processes. S2ESC recommends that organizations should use 12207 and related standards to assist them in defining their organizational processes and then assess those processes using ISO 9001 or alternative methods. However, this book is based on the premise that the reader's processes may not have been defined in this fashion and that the reader seeks help in improving those processes in anticipation of an ISO 9001 assessment. Therefore, the advice provided in this book does not make the assumption that the user's processes conform to 12207. In general, that advice will be consistent with 12207, but its intent is to assist with process documentation and improvement in the context of ISO 9001 compliance. It is the premise of the information provided within these pages that IEEE software engineering standards [41] can be used to provide the basic beginning framework for this type of process improvement.

IEEE standards can be used as tools to help with the process definition and documentation (e.g., work products) required in support of the process improvement of software and systems development efforts. Many of the IEEE software engineering standards provide detailed procedural explanations, they offer section-by-section guidance on building the necessary support material and, most importantly, they provide best-practice guidance in support of process definition as described by those from academia and industry who sit on the panels of standards reviews.

ISO 9001 does not tell the user "how" to satisfy its conformance criteria. ISO 9001 is a general quality standard with a process-based management approach, describing the criteria that the processes should support. IEEE standards are prescriptive. These standards describe how to fulfill the requirements associated with all the activities of an effective software and systems project life cycle.

IEEE standards are highly specific. They are documented agreements containing technical specifications or other precise criteria to be used consistently as rules, guidelines, or definitions of characteristics to ensure that materials, products, processes, and services are fit for their purpose. In contrast, ISO 9001 is a generic quality-management-system standard. "Generic" means that the same standards can be applied to any organization, large or small, whatever its product or service, in any sector of activity, and whether it is a business enterprise, a public administration, or a government department. "Management system" refers to what the organization does to manage its processes or activities in order that the products or services that it produces meet the objectives it has set itself. "Management system standards" provide the organization with a model to follow in setting up and operating the management system. "Quality management" is what the organization does to ensure that its products or services satisfy the customer's quality requirements and comply with any regulations applicable to those products or services. ISO 9001 concerns the way an organization goes about its work, and is not directly concerned with the result of this work. ISO 9001 has to do with processes, not products (at least not directly). However the way the organization manages its processes is going to affect its final product. The efficient and effective management of processes is going to affect whether or not everything has been done to ensure that the product satisfies the customer's quality requirements.

It is often hard to separate the details associated with product development from the practices required to manage the effort. Simply handing the ISO 9001 standard to a project lead or manager provides them with a description of an end results model. Pairing this with IEEE standards provides them with a way to work toward this desired end. IEEE standards do not offer a "cookie cutter" approach to management; rather, they support the definition of the management processes in use by describing what is required.

Organizations are motivated to use ISO 9001 from two directions: external and internal. External motivations are your customers, investors, or competitors. Your customers may ask questions in their request for quote (RFQ). ISO 9001 may show up in their proposed contracts or requirements. Customer inspections or second-party audits may use ISO 9001. Internal motivations for ISO 9001 are needs for improvement, expected growth, or part of a change management strategy. ISO 9001 improvements have resulted in lowered costs, reduced time cycles, improved product quality, and increased resources for new product features. External motivations normally lead to registration to ISO 9001 using third-party auditors, whereas internal motivations lead to self-declaration of conformance to ISO 9001. Thus conformity to the ISO 9001 requirements may be verified by self-declaration, second-party, or third-party registration.

Many software engineering organizations have implemented ISO 9001 in a minimal form to the level of self-declaration or registration, but have not realized its full potential. This is due to the general and abstract nature of ISO 9001, contrasted by the inherent complexity of software engineering processes. Successful software engineering organizations create well-formed internal work products or artifacts during development phases leading to a finished product. Although ISO 9001 does not describe these artifacts, this book will describe and provide examples of these important artifacts that are aligned with the IEEE software engineering standards. Conformance to ISO 9001 using artifacts related to well-known IEEE standards provides benefits of increased confidence to knowledgeable customers/investors and increased clarity to project personnel. The usage of IEEE-related artifacts affords customers/investors additional opportunities for interaction during the various project phases leading to refined requirements, feedback, and expectations. The increased clarity among project personnel promotes easier personnel movement between teams and organizations, better alignment between project/program management and software engineering, and better alignment between company goals, objectives, and project/program success factors.

For organizations that do not wish to pursue ISO 9001 registration, this text will show how the application of IEEE standards, and their use as reference material, can facilitate the development of sound software and systems engineering practices. This book is geared for the ISO 9001 novice, the project manager, and software engineering practitioners who want a one-stop source, helpful document, that provides the details and implementation support required when pursuing process definition and improvement. The plans described in this book may be used as an integrated set in support of software and systems process definition and improvement. It should be noted that many of the plans offer examples, these examples are meant to show intent and offer only narrow perspectives on possible plan content.

The authors recommend attendance both in an ISO 9001:2000 Implementation Course (about 2 days) and acquisition of the IEEE Software Engineering Collection that fully integrates over 40 of the most current IEEE software engineering standards onto one CD-ROM. This book provides a mapping between the IEEE standards set. This mapping is available in Appendix D.

What is ISO 9001?

The history of ISO 9001 starts with the 1987 international version, based largely on the British version. It was revised in 1994 and completely rewritten into a process approach in 2000, thus labeled ISO 9001:2000 [47]. ISO 9001 has been touted as the world's most successful standard. The worldwide total of certificates to the ISO 9001:2000 quality management systems standard at the end of 2004 was 670, 399. Certificates had been issued in 154 countries. ISO 9001 is a necessary requirement for a quality management system. It is a part of ISO 9000 family that consist of ISO 9000 (fundamentals and vocabulary), ISO 9001 (requirements), ISO 9004 (guidelines for performance improvements) [49], and ISO 19011 (guidelines for quality and environmental management systems auditing) [56]. ISO 9001 is based on eight fundamental quality principles from ISO 9000:

1. Customer focus
2. Leadership
3. Involvement of people
4. Process approach
5. System approach to management
6. Continual improvement
7. Factual approach to decision making
8. Mutually beneficial supplier relationships

All of these principles relate to successful software engineering practices.

ISO 9001 represents an international consensus on good management practices with the aim of ensuring that the organization can time and time again deliver the products or services that meet the customer's quality requirements, and applicable regulatory requirements, while aiming to enhance customer satisfaction and achieve continual improvement of its performance in pursuit of these objectives.

ISO 9001 is a generic quality management standard with a process-based management approach, having over 100 "shalls." In business-to-business transactions, the customers and investors have some level of confidence that the organization is at least delivering what is expected and that organizational improvements are underway for better products in the future, as ISO 9001 requires that processes to be continually improved even after achieving ISO conformance.

As a generic standard, written for use by the widest possible audience, it has been adopted by various industry sectors (e.g., telecom, aviation, automobile) and aligned with other models and frameworks. This book will describe the alignment to IEEE software engineering standards and associated artifacts.

What ISO 9001 is Not

ISO 9001 does not

- Ensure that a product or service is without defects
- Provide specific instructions on how to build the quality system
- Provide for easy interpretation and, thus, implementation

- Provide a rulebook, prescribe formats, prescribe explicit contents
- Prescribe media or specify technology
- Specify requirements for the products or services
- Specify all of the product realization processes
- Offer details about its application to specific domains of expertise
- Prescribe measurement definitions
- Focus on associated financial issues

ISO 9001 does not provide processes or process descriptions. Actual processes are dependent upon:

- Application domain(s)
- Organization structure
- Organization size
- Organization culture
- Customer requirements or constraints

This book will provide the guidance on details about its application to the software engineering domains of expertise as a management tool to attest to the software engineering process and how it will be managed and reviewed.

What are Standards?

Standards are documented agreements on a set of rules and guidelines that provide a common framework for communication and are expressed expectations for the performance of work. These rules and guidelines set out what are widely accepted as good principles or practices in a given area. They provide a basis for determining consistent and acceptable minimum levels of quality, performance, safety (low risk), and reliability.

Standards are consensus-based documents that codify best practice [42]. Consensus-based standards have seven essential attributes that aid in process engineering. They

1. Represent the collected experience of others who have been down the same road
2. Describe in detail what it means to perform a certain activity
3. Can be attached to or referenced by contracts
4. Help to assure that two parties have the same meaning for an engineering activity
5. Increase professional discipline
6. Protect the business and the buyer
7. Improve the product

IEEE software engineering standards provide a framework for defining and documenting software and systems engineering activities. The structure of the standards set supports the criterion as described in ISO 9001. The structure of the IEEE Software Engineering Standards Set provides for organizational adaptation. Each standard describes recommended best practices detailing required activities. These standards documents provide a common basis for documenting organizationally unique process activities.

Chapter **2**

Summary of ISO 9001

ISO 9001 PRINCIPLES

ISO 9001 is based on the following the eight Quality Management Principles that can be used by management to lead an organization toward improved performance:

1. *Customer focus.* Organizations depend on their customers and therefore should understand current and future customer needs, meet customer requirements, and strive to exceed customer expectations. For software engineering organizations, this involves requirements analysis and architectural design processes to acquire and translate customers needs and expectations into software requirements.

2. *Leadership.* Leaders establish unity of purpose and direction of the organization. They should create and maintain the internal environment in which people can become fully involved in achieving the organization's objectives. For software engineering organizations, this translates into informed leaders coaching their teams toward visible targets and allowing for team members to make decisions supporting the overall direction. Management supports the chosen methodology and the management and technical reviews of plans and deliverables.

3. *Involvement of people.* People at all levels are the essence of an organization and their full involvement enables their abilities to be used for the organization's benefit. The chosen methodology is aligned with the project plan, which defines everyone's roles and accountabilities, including reviews at all levels and phases, where people freely share their knowledge and experience.

4. *Process approach.* A desired result is achieved more efficiently when activities and related resources are managed as a process. Management supports the chosen methodology, which is sufficiently and systematically defined as processes, activities, tasks, and outcomes for all roles and deliverables. Risks and defects are expected, discovered, analyzed, and corrected without blame. Management provides key staff with process management training. Processes are managed and improved for reduction of defects and cycle time.

5. *System approach to management.* Identifying, understanding and managing

interrelated processes as a system contributes to the organization's effectiveness and efficiency in achieving its objectives. The chosen SE methodology is systematically defined as processes, activities, tasks, and outcomes for both organization's goals and customers' needs, while understanding the interdependencies between the processes of the system.

6. *Continual improvement.* Continual improvement of the organization's overall performance should be a permanent objective of the organization. Management provides people with training in the methods and tools of continual improvement, in order to work with a consistent organization-wide approach to reach management goals. The benefits are discussion later in this chapter.

7. *Factual approach to decision making.* Effective decisions are based on the analysis of data and information. Management provides process measurement infrastructure, which is accessible and analyzed for subsequent process changes and inputs to management reviews.

8. *Mutually beneficial supplier relationships.* An organization and its suppliers are interdependent and a mutually beneficial relationship enhances the ability of both to create value.

Software engineering organizations can be viewed by their customers as key suppliers who provide tools or software subsystems for inclusion to sell, or larger systems for tailoring. Key suppliers are selected, and their relationships cultivated through clear and open communication and planning. Suppliers are inspired, encouraged, and recognized for their improvements and achievements. ISO 9001 principles can lead a supplier to both operational improvement and to increased customer satisfaction and industry recognition.

Why Should My Organization Implement ISO 9001?

ISO 9000 provides a tried and tested framework for taking a systematic approach to managing your business processes so that they consistently turn out products and services conforming to the customer's expectations.

The ISO 9001 focus is on business objectives. The standard is intended to deliver measurable quality objectives based on a quality policy. However, to be effective, the quality policy must be consistent with the plans and strategy of the business. ISO 9001 approach can offer real strategic value, especially if it is aligned with corporate goals and objectives. Achievement of an organization's business goals occurs when ISO 9001 implementation is designed to align quality and business objectives, uses process management techniques, and integrates continual improvement activities into the business plans.

How Does the ISO 9001 Model Work?

The requirements for a quality system have been standardized, but most of us like to think our business is unique. So how does ISO 9000 allow for the diversity of say, on the one hand, a "Mr. and Mrs." enterprise, and on the other, a multinational manufacturing company with service components, or a public utility, or a government administration?

The answer is that ISO 9000 lays down what requirements your quality system must meet, but does not dictate how they should be met in your organization, which allows great scope and flexibility for implementation in different business sectors and business cultures, as well as different national cultures.

What If My Organization Implements ISO 9001?

Implementation details are covered later in this book. As a minimum, it consists of conformance and usage of ISO 9001. Usage includes records and audits of conformance, followed by management reviews and corrective and preventative actions recorded and completed. Your organization can implement and benefit from ISO 9001 without seeking to have your management system audited and certified as conforming to the standards by an independent, external certification body. Your organization can implement the voluntary standard ISO 9001 solely for the internal benefits it brings in increased effectiveness and efficiency of your operations, without incurring the investment required in a certification program. Your organization would self-declare conformance to ISO 9001.

Another course of action is deciding to have an independent audit of your system to confirm that it conforms to the standard. This is a decision to be taken on business grounds, such as:

- A contractual, regulatory, or market requirement
- Meets customer preferences
- Part of a risk management program
- Motivate your staff by setting a clear goal for the development of the management system

ISO 9001 Audits

The organization should itself audit its ISO 9001-based quality system to verify that it is managing its processes effectively and that it is fully in control of its activities.

In addition, the organization may invite its clients to audit the quality system in order to give them confidence that the organization is capable of delivering products or services that will meet their requirements.

Lastly, the organization may engage the services of an independent quality system certification body to obtain the ISO 9001 certificate of conformity. This last option has proved extremely popular in the marketplace because of the perceived credibility of an independent assessment. The organization may thus avoid multiple audits by its clients, or reduce the frequency or duration of client audits. The certificate can also serve as a business reference between the organization and potential clients, especially when supplier and client are new to each other, or far removed geographically, as in an export context.

ISO 9001 Conformance, Registration, and Accreditation

As a minimum, organizations can self-declare "conformance" to ISO 9001 as follows:

"Certification" refers to the issuing of written assurance (the certificate) by an independent, external body that has audited an organization's management system and verified that it conforms to the requirements specified in the standard.

"Registration" means that the auditing body then records the certification in its client register. "Certification" seems to be the term most widely used worldwide, although registration (in which "registrar" is used as an alternative to registration/certification body) is often preferred in North America, and the two are also used interchangeably.

"Accreditation" refers to the formal recognition by a specialized body—an accreditation body—that a certification body is competent to carry out ISO 9001 certification in specified business sectors. Compliant organizations are not accredited; only external bodies (called registrars) can be accredited.

Basic Business Model for Software Engineering Organizations

This book assumes your software engineering (SE) organization is either internally supporting the company or externally providing software products and services to your customers. In the first case, your SE organization maybe a small part of the company; an example is the Information Technology (IT) Group supporting the company's IT structure. In the second case, the SE organization may be the most important to the company mission.

In either case, the company's strategic plan or annual business plan should identify directions, objectives, and risks for the SE organization. Many companies use strategic planning for the start of their risk planning and management. The Human Resource (HR) Department has several responsibilities impacting the SE organization, including job descriptions, staffing goals, training, and recognition plans. After plans and resources, portfolio management of the SE projects allows smooth startup, transition, and ending of projects. The SE life cycle is usually reviewed by portfolio managers at important project phases. The focus is on the project success criteria. These reviews are usually key points in the risk management at both the business and the project levels. Project planning starts with software development life cycle (SDLC) selection and tailoring, followed by project plans and project management of the SDLC processes and roles. The project members work their roles within the processes and create, review, revise, and transmit project artifacts to other project members. Project artifacts are stored and managed as part of the project configuration management system. The final artifacts are customer deliverables, installations, and/or training.

Conformance Pathways

When the scope of the company's activities includes areas other than SE, the relationship between the SE elements of that organization's quality management system and the remaining aspects should be clearly documented within the quality management system as a whole. ISO 9001 Clauses 4, 5, and 6 and parts of Clause 8 are applied mainly at the "enterprise" level in the organization, although they do have some effect at the "project/product level." Each SE project or product development may tailor the associated parts of the organization's quality management system, to suit project/product-specific requirements.

This freedom of tailoring has resulted in many instances of ineffective SE management systems. Some were implemented only to quickly get a certificate to show to a demanding customer, but later resulted in retrenchment due to inefficiencies and lack of compliance, and then finally loss of the certificate. Other implementations have resulted in neutered systems without any improvement levers and lack of compliance to the spirit of ISO 9001. This book will help you avoid those dead-end paths.

This book will guide your implementation of the ISO 9001 requirements, following its flexibility for framework, life cycle, and process descriptions. Specifically, this book will follow the IEEE 12207 SDLC standard in processes description as aligned by IEEE 90003 Software Guidance for ISO 9001. Furthermore, this book will describe typical project artifacts or outcomes that are aligned with specific IEEE SE standards. (Outcomes are observable results and provide more specific indications of what the process, activity, or task should accomplish when successfully enacted.) This approach and structure allows for both ISO 9001 audits and ISO/IEC 15504 [55] process assessments.

ISO 9001 Benefits

ISO 9001 Implementation for Improvement

When companies choose ISO 9001 implementation for improvement, there are many benefits:

Improved customer satisfaction due to:
- Fewer customers found errors.
- New features or systems delivered on time with agreed upon content.
- Increased perceived employee responsiveness to customer needs and requirements.

Increased profits due to:

- Decreased development expenses due to rework resulting from less defects created within the SDLC.
- Less expenses with fewer schedule overruns.
- Increased productivity as the correct data, equipment, tools, and documentation are integrated into the total process.
- Lower after-sales support costs.
- Breakdown of internal organizational boundaries as staff from different functions and levels work together.
- Elevated confidence among sales force as quality improves.
- Elevated confidence among employees as understanding of how ISO efforts and your business strategy are aligned for organizational improvement.
- Confident managers addressing problems with long-term impact by focusing on the quality management system continual improvement processes.

ISO 9001 Implementation for Customer and Competitor Forces

When companies react to customer and competitor forces by choosing ISO 9001 implementation, they derive several benefits, such as enhanced marketplace recognition due to public, independent certification.

Certification can also improve customer and investor confidence due to planned interactions during the SDLC phases:

- Concept and definition
- Requirements
- Test planning
- Early demonstrations/simulations
- Predelivery measurements shared
- Postdelivery responsiveness

Benefits of the Process Approach

Improvements in customer and investor satisfaction, revenue, expenses, and employee morale are managed through the continual improvement processes that focus on key business and SE processes:

- Integration and alignment of processes to enable achievement of planned results
- Ability to focus effort on process effectiveness and efficiency
- Provision of confidence to customers, and other interested parties, about the consistent performance of the organization

- Transparency of operations within the organization
- Lower costs and shorter cycle times, through the effective use of resources
- Improved, consistent, and predictable results
- Provision of opportunities for focused and prioritized improvement initiatives
- Encouragement of the involvement of people and the clarification of their responsibilities

Chapter 3

Relationship to Software Engineering Standards

STANDARDS ORGANIZATIONS

Standards organizations are bodies, organizations, and institutions that produce standards. These organizations develop standards to provide stability and consistency, with the hope of lowering costs in any industry or enterprise. Software standards play an important role as they can be used to reduce costs and complexity when buying software systems and they can be used to monitor the quality of the systems and products that are produced. Software standards are also an excellent reference on what is considered good practice by the international community of professionals that work in these areas.

ISO Technical Committee 176 on Quality Management and Quality Assurance

The International Organization for Standardization (ISO) is a network of the national standards institutes of over 150 countries, formed on the basis of one member per country. The United States is represented by the American National Standards Institute (ANSI). The hallmark of ISO standards are that they are consensus-based, market driven, voluntary, and are used in worldwide agreements. ISO has numerous Technical Committees (TCs). TC 176 on Quality Management and Quality Assurance is responsible for developing the ISO 9000 series of standards and guidance documents. The TC 176 scope is standardization in the field of generic quality management, including quality systems, quality assurance, and generic supporting technologies, including standards that provide guidance on the selection and use of these standards. Member nations maintain and revised the Quality Management and

Quality Assurance Standards. TC 176 produced the ISO 9001:2000 in December 2000 and expects its revisions to be published by year end 2008.

International Electrotechnical Commission

The International Electrotechnical Commission (IEC) is the leading global organization that prepares and publishes international standards for all electrical, electronic, and related technologies. Together, ISO, IEC, and ITU (International Telecommunication Union) have built a strategic partnership with the WTO (World Trade Organization) with the common goal of promoting a free and fair global trading system.

ISO/IEC Joint Technical Committee 001

For software and system standards, ISO and IEC created the ISO/IEC Joint Technical Committee 001 "Information technology" (JTC1) for standardization in the field of information technology. JTC1 Information Technology includes the specification, design, and development of systems and tools dealing with the capture, representation, processing, security, transfer, interchange, presentation, management, organization, storage, and retrieval of information.

ISO/IEC JTC1 SC7 Software and Systems Engineering Standards Committee

One JTC1 subcommittee (SC) is SC7 Software and Systems Engineering. This SC was created in 1997, with the scope of the standardization of processes, supporting tools, and supporting technologies for the engineering of software products and systems. Figure 3-1 illustrates how SC7 scope interacts with other SCs and disciplines.

ISO TC176 transferred ISO 9001 software guidance responsibilities to ISO/IEC JTC1/SC7. Thus, JTC1/SC7 published the ISO/IEC 90003 *Software and System Engineering—Guidelines for the Application of ISO 9001:2000 to Computer Software* in 2004. JTC1/SC7 has published almost 90 standards and 25 are currently under development. Due to the large collection of software engineering standards developed by the IEEE Software and Systems Engineering Standards Committee, JTC1/SC7 has an active liaison with IEEE Computer Society.

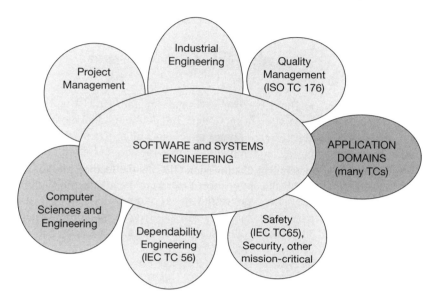

Figure 3-1. Interaction of SC7 scope with other TCs and disciplines.

American National Standards Institute

The American National Standards Institute (ANSI) is a private, nonprofit organization [501(c)3] that administers and coordinates the U.S. voluntary standardization and conformity assessment system. The Institute's mission is to enhance both the global competitiveness of U.S. business and the U.S. quality of life by promoting and facilitating voluntary consensus standards and conformity assessment systems, and safeguarding their integrity.[1] ANSI facilitates the development of American National Standards (ANS) by accrediting the procedures of standards-developing organizations (SDOs). These groups work cooperatively to develop voluntary national consensus standards. Accreditation by ANSI signifies that the procedures used by the standards body in connection with the development of American National Standards meet the Institute's essential requirements for openness, balance, consensus, and due process. ANSI has over 200 SDOs, with the 20 largest SDO producing 90% of the standards. One of the largest SDO is IEEE-SA.

---

[1]www.ansi.org

Institute of Electrical and Electronics Engineers

The Institute of Electrical and Electronics Engineers, Inc. (IEEE) is a nonprofit technical professional association of more than 360,000 individual members in approximately 150 countries. IEEE promotes the engineering process of creating, developing, integrating, sharing, and applying knowledge about electronic and information technologies and sciences for the benefit of humanity and the profession. The Institute provides groups the opportunity to take part in shaping the direction of technology and its marketplace application by developing industry-driven standards. IEEE holds annually more than 300 major conferences and produces 30% of the world's published literature in electrical engineering, computers, and control technology.[2]

The Institute provides wide technology coverage over 40 societies. The largest is the IEEE Computer Society (IEEE CS). The IEEE CS is a leading international provider of technical information and services in computer and information processing technology and was founded in 1946. Today, with nearly 100,000 members, the IEEE CS is the world's leading organization of computer professionals. IEEE CS is dedicated to advancing the theory, practice, and application of computer and information processing technology. The Society is considered to be the leading provider of technical information and services to the world's computing professionals. It fosters international communication, cooperation, and information exchange, as over 40% of the members live and work outside the United States.[3]

The IEEE SDO is the Standards Association (IEEE-SA). The IEEE-SA is a world-renowned, independent American National Standards Institute (ANSI) accredited SDO. The IEEE-SA is an independent professional and credible standards-setting body that develops industry-driven standards based on current scientific consensus, either through individual experts or corporate entities. IEEE-SA is an honest information broker for resolving issues and developing consensus to find industry solutions. They provide benefits for corporate growth and exposure, and to help provide business initiatives with lower risk through use of IEEE standards. IEEE-SA maintains a large portfolio of more than 900 active standards and 700 in-development projects. These standards and development projects are managed by over 140 Standards Committees. One such Standards Committee is the Software and Systems Engineering Standards Committee (S2ESC). IEEE standards are internationally implemented in an array of environments, all resulting in the improved competitiveness of companies incorporating them and adding value to industry and the marketplace.

[2]www.ieee.org
[3]www.computer.org

IEEE S2ESC Software and Systems Engineering Standards Committee

The IEEE Software and Systems Engineering Standards Committee (S2ESC) is rapidly approaching its 30th anniversary. In 1976, this arm of the IEEE Computer Society was chartered to develop the first standards for software engineering. S2ESC was chartered with the vision to develop a family of products and services based on software engineering standards for use by practitioners, organizations, and educators to: (1) improve the effectiveness and efficiency of their software engineering processes; (2) improve communications between acquirers and suppliers, and (3) to improve the quality of delivered software and systems containing software. This section describes S2ESC, the contribution it has made to the software engineering community over these last 30 years, and how S2ESC continues to ensure that its products are relevant, valued, and reflect best current practice [69].

From a narrow perspective, S2ESC manages the scope and direction of IEEE Software and Systems Engineering and Standards. S2ESC is the standards arm of the IEEE Technical Council on Software Engineering (TCSE), and has worked to provide a standards collection that:

- Provides a consistent view of the state of the practice
- Is aligned with the Software Engineering Body of Knowledge (SWEBOK)
- Addresses practitioner concerns
- Is affordable

From a broader perspective, in addition to the development of standards, S2ESC develops supporting knowledge products and sponsors or cooperates in annual conferences and workshops in its subject area. S2ESC also participates in international standards making as a member of the U.S. Technical Advisory Group (TAG) to ISO/IEC JTC1/SC7 and as a direct liaison to SC7 itself.

The mission of the S2ESC is:

1. To develop and maintain a family of software and systems engineering standards that is relevant, coherent, comprehensive, and effective in use. These standards are for use by practitioners, organizations, and educators to improve the effectiveness and efficiency of their software engineering processes, to improve communications between acquirers and suppliers, and to improve the quality of delivered software and systems containing software.

2. To develop supporting knowledge products that aid practitioners, organizations, and educators in understanding and applying our standards.

3. To support and promote a Software Engineering Body of Knowledge, certification mechanisms for software engineering professionals, and other products contributing to the profession of software engineering.

ISO/IEC JTC1 SC7 Software and Systems Engineering Standards Committee (S2ESC)

At the center of software engineering standardization in the United States is the Software and Systems Engineering Standards Committee (S2ESC) of the IEEE Computer Society. This IEEE CS standards development sponsor maintains over 40 standards that directly support the practice of software and systems engineering. The counterpart of S2ESC in the international forum is ISO/IEC JTC1 SC7 S2ESC participates in SC7 through the Computer Society's membership in the U.S. Technical Advisory Group (TAG) that formulates national positions and selects the delegation for meetings of SC7. The CS also has a category A liaison to SC7, permitting direct participation with the exception of voting. Figure 3-2 shows the relationship of the IEEE CS to other SDOs.

RELATIONSHIPS AMONG ISO 9001, ISO 90003, IEEE 12207, AND ISO/IEC 15504

There are relationships among quality assurance, life cycle processes, and process assessment standardizations. ISO 9001, ISO/IEC 90003, IEEE 12207, and ISO/IEC 15504 are all important to completely understand the requirements of ISO 9001 for a software engineering organization. ISO 9001 represents quality assurance at the system level. ISO/IEC 90003 gives the SE organization guidance on how to apply ISO 9001 on a software project. IEEE 12207 represents the processes employed throughout the life cycle of a software product. Each one of the ISO/IEC 90003

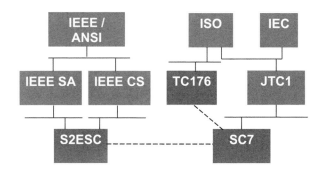

Figure 3-2. Relationship of S2ESC to other SDOs.

clauses that are software unique point to the applicable process in IEEE 12207 software development life cycle to provide more details on implementation considerations. ISO/IEC 15504 represents process assessment as applied in SE organizations. IEEE 12207 provides the baselines of life cycle processes to the ISO/IEC 15504 assessment process. ISO 9001 provides the basis for quality assurance to both IEEE 12207 and ISO/IEC 15504 as described by Figure 3-3.

SOFTWARE ENGINEERING BODY OF KNOWLEDGE (SWEBOK)

The purpose of the Guide to the Software Engineering Body of Knowledge (SWEBOK) is to provide a consensually validated characterization of the bounds of the software engineering discipline and to provide a topical access to the Body of Knowledge supporting that discipline.

Software Quality is one of the ten SWEBOK Knowledge Areas (KA) and deals with software quality considerations that transcend the software life cycle processes. As software quality is a ubiquitous concern in software engineering, it is also considered in many of the other SWEBOK KAs, including Software Engineering Management and Software Engineering Process. The Software Engineering Management KA can be defined as the application of management

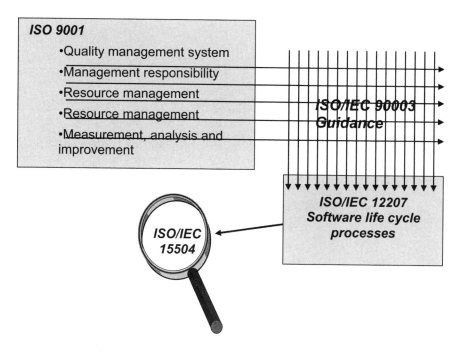

Figure 3-3. ISO 9001 as a basis for quality assurance.

activities—planning, coordinating, measuring, monitoring, controlling, and reporting—to ensure that the development and maintenance of software is systematic, disciplined, and quantified. The Software Engineering Process KA is concerned with the definition, implementation, assessment, measurement, management, change, and improvement of the software life cycle processes themselves. ISO 9001 implementation and usage will directly contribute to these three SWEBOK KAs.

In order to circumscribe software engineering, it is necessary to identify the disciplines with which software engineering shares a common boundary. Two of the eight SWEBOK Related Disciplines of Software Engineering cover the usage of ISO 9001 as described by Table 3-1.

Quality Management is one of the eight Related Disciplines and is defined in ISO 9000-2000 as "coordinated activities to direct and control an organization with regard to quality." The main reference on quality management is ISO 9001:2000 Quality Management Systems Requirements. Project Management (PM) is another Related Disciplines. Project Management is defined in the Guide to the Project Management Body of Knowledge" (PMBOK Guide). PMBOK has nine Knowledge Areas, one is Project Quality Management, which is intended to be compatible with that of the ISO 9001 and has three major project quality management processes: Quality Planning, Quality Assurance, and Quality Control.

CAPABILITY MATURITY MODEL INTEGRATED (CMMI)

The Software Engineering Institute (SEI) developed the Capability Maturity Model Integrated (CMMI) framework in 2002. Portions of this framework are described as

Table 3-1. SWEBOK relationship to ISO 9001

Knowledge areas	Topical areas	Topics related to ISO 9001
Software Quality	Software Quality Fundamentals	Software Engineering Process Quality
		Quality Improvement
	Software Quality Management Processes	Software Quality Assurance
	Practical Considerations	Software Quality ManagementTechniques
Software Engineering Management	Software Project Planning	Quality Management
Software Engineering Process	Process Definition	Software Life Cycle Processes
	Process Assessment	Process Assessment Models

generic practices (GP) and are applicable to all CMMI process areas. Generic practices provide institutionalization to ensure that the processes associated with the process area will be effective, repeatable, and lasting. In a similar purpose, ISO 9001 requires institutionalization to ensure that all business processes will be effective, repeatable, and lasting. Table 3-2 provides a cross-reference of CMMI GP to ISO 9001 requirement clauses and shows good coverage for most of ISO 9001.

Table 3-2. CMMI generic practices and ISO 9001 cross reference

Abbreviation	CMMI generic goals and practices	ISO 9001 clauses
GG 1	**Achieve Specific Goals**	
GP 1.1	Perform base practices.	6.2.1
GG2	**Institutionalize a Managed Process**	
GP 2.1	Establish and maintain an organizational policy for planning and performing the processes.	4.1, 4.2.1, 5.1, 5.5.3, 7.6, 8.2.2
GP 2.2	Establish and maintain the plan for performing the processes.	4.1, 4.2.2, 5.4.2, 7.1, 7.3, 7.5.1, 7.6, 8.1, 8.2.3
GP 2.3	Provide adequate resources for performing the processes, developing the work products, and providing the services of the process.	4.1, 6.1, 7.5.1
GP 2.4	Assign responsibility and authority for performing the process, developing the work products, and providing the services of the processes.	5.5.1, 8.2.2
GP 2.5	Train the people performing or supporting the processes as needed.	6.2.1
GP 2.6	Place designated work products of the processes under appropriate levels of configuration management.	4.1, 4.2.3, 4.2.4, 7.3.7, 7.5.1
GP 2.7	Identify and involve the relevant stakeholders of the processes as planned.	5.1, 7.2.3, 7.3.2, 7.3.4
GP 2.8	Monitor and control the processes against the plan for performing the process and take appropriate corrective action.	4.1, 7.5.1, 7.6, 8.2.3
GP 2.9	Objectively evaluate adherence of the processes against its process description, standards, and procedures, and address noncompliance.	4.1, 7.6, 8.2.2
GP 2.10	Review the activities, status, and results of the processes with higher-level management and resolve issues.	5.6.1, 5.6.2, 5.6.3, 7.2.2, 7.3.2, 7.6
GG3	**Institutionalize a Defined Process**	
GP 3.1	Establish a defined process.	5.4.2, 7.1
GP 3.2	Collect improvement information.	8.4

Chapter 4

Implementation Guidance

IMPROVEMENT FRAMEWORKS SELECTION

Plan, Do, Check, Act (PDCA) Cycle

W. Edwards Deming in the 1950s proposed that business processes should be analyzed and measured to identify sources of variations that cause products to deviate from customer requirements. He recommended that business processes be placed in a continual feedback loop so that managers can identify and change the parts of the process that need improvements. Deming created a simple diagram to illustrate this continual process, commonly known as the PDCA cycle for Plan, Do, Check, Act:

PLAN: Design or revise business process components to improve results.
DO: Implement the plan and measure its performance.
CHECK: Assess the measurements and report the results to decision makers.
ACT: Decide on changes needed to improve the process.

The PDCA is endorsed by ISO 9001 in paragraph "0.2 Process Approach" [47]. This PDCA cycle was originally developed by Walter A, Shewhart, a Bell Laboratories scientist, who was Deming's friend and mentor, and the developer of statistical process control (SPC) in the late 1920s. So sometimes the PDCA is referred to as the "Shewhart Cycle."

IDEAL (Initiating, Diagnosing, Establishing, Acting, and Learning) Model

As PDCA is a simple generic improvement model, the Software Engineering Institute (SEI) originally developed IDEAL to support the SW-CMM®-based software process improvement model [96]. This improvement model can serve to lay the groundwork for a successful improvement effort (initiate), determine where an organization is in reference to where it may want to be (diagnose), plan the specifics of how to reach goals (establish), define a work plan (act), and apply the lessons

Practical Support for ISO 9001 Software Project Documentation. By S. Land and J. Walz

learned from past experience to improve future efforts (learn). This robust improvement cycle has five phases, each with fourteen activities, with "Stimulus for Change" starting up IDEAL (see Figure 4-1):

- Initiating
  - ○ Set Context
  - ○ Build Sponsorship
  - ○ Charter Infrastructure
- Diagnosing
  - ○ Characterize Current and Desired States
  - ○ Develop Recommendations
- Establishing
  - ○ Set Priorities
  - ○ Develop Approach
  - ○ Plan Actions
- Acting
  - ○ Create Solution
  - ○ Pilot/Test Solution
  - ○ Refine Solution
  - ○ Implement Solution
- Learning
  - ○ Analyze and Validate
  - ○ Propose Future Actions

IDEAL can be mapped to PDCA, where P starts at Charter Infrastructure and ends with Create Solution, D is Pilot/Test Solution, C is Refine Solution, and A starts with Implement Solution. This chapter will use the IDEAL model for initial ISO 9001 implementation and ongoing maintenance.

SET CONTEXT

It is important to understand the context driving change. Often, the dominate force for change is in support of business improvement or customer interest. These two forces for ISO 9001 implementation for conformity—business improvement and/or customer or investor confidence—require focusing on different issues. The former requires more change activities with a focus on return on investment (ROI). The latter requires a focus on minor process changes for alignment, consolidation, training, and documentation. If management has determined that there is a lack of customer or investor confidence, then business improvement is often the stronger force setting the context for change.

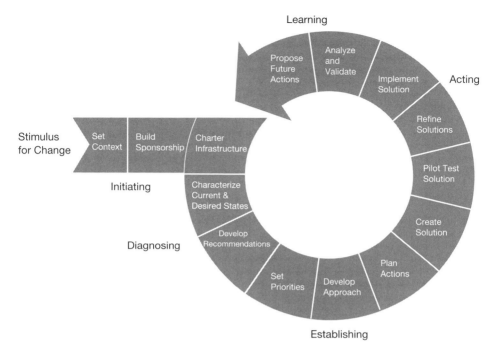

Figure 4-1. The IDEAL model.

Business Improvement through ISO 9001 Implementation

This chapter will assume that business improvement is the main driver or force for change. Implementation for conformity to ISO 9001 is a project whose long-term maintenance to conformity is actually a process. The ISO 9001 implementation project and processes are different and separate from the Software Engineering (SE) projects and processes. Several approaches to the initial business improvement project and its ongoing maintenance process will be described.

Customer and Investor Confidence through ISO 9001 Implementation

Although customers and investors are normally pleased that management has committed to a formal ISO 9001 implementation, they more engrossed in management's commitments for existing SE products and projects. The selected software development life-cycle model, which will be discussed later, should offer several interfaces to the customer or investor that can be confidence builders and manage expectations. These interfaces include:

- Concept presentation
- Requirements gathering and management
- Feature prioritization
- Acceptance test planning use cases
- Early demonstrations/simulations and feedback
- Predelivery measurements
- Postdelivery error reporting and fixing responsiveness
- Change control priorities and approval

BUILD SPONSORSHIP

Consensus and agreement in support of improvement must be developed. This requires a gathering together of all stakeholders and the definition of their roles. Sometimes, this requires executive-level training on ISO 9001 and IEEE SE standards.

Improvement Project Stakeholders

For the business improvement project to implement ISO 9001, the stakeholders need to be identified and committed to their roles. Several stakeholders have been identified as follows:

- **Senior management** initiates the implementation/improvement project, commits to the investment, monitors the various stages toward completion, and guides the organization through is periodic review and planning cycles.
- **Customers, investors,** or their representatives have a variety interfacing roles, depending on their maturity, previous experience, and type of contracting agreements.
- **Software engineering team members** have technical and supporting roles as defined in their business improvement model and their software development life cycle model. The roles are often aligned with technical process requirements. Chapters 5, 6, and 7 will define these processes.

CHARTER INFRASTRUCTURE

The definition of the ISO 9001 implementation project much be accomplished. The project should be described in terms of time frame, membership, organizational and project scope, and level of conformity (self-declared or external assessment). The degree of formality, required infrastructure investments, and communication channels should also be defined.

Establish Steering Committee and Process Group

Process improvement normally works best with two levels of committees or groups. The first is the Steering Committee, which includes the sponsors and meets infrequently through the year. The second is the Process Group, a process improvement team comprising a chosen group of people who are given responsibility and authority for improving a selected process in an organization; this team must have the backing of senior management represented by the Steering Committee. Process owners in the Process Group are responsible for the process design, not for the performance, of their associated process areas. The process owner is further responsible for the process measurement and feedback systems, the process documentation, and the training of the process performers in its structure and conduct. In essence, the process owner is the person ultimately responsible for improving a process. IEEE software engineering standards provide valuable support to the process team and each individual process owner. The standards can be used to help define and document the initial baseline of recommended processes and practices.

Software Engineering Training

Process Group members need SE skills. Implementing process improvement can be very time-consuming, depending upon the scope and complexity of the process. Expectations for the process owner's time commitments and job responsibilities must be modified accordingly to reflect the new responsibilities. This commitment should reflect time budgeted for process definition and improvement and any required refresher training. Traditionally, individuals graduating with computer science degrees have not been trained in a manner meeting this definition. In fact, except for universities that offer undergraduate or masters of software engineering courses, computer science students have been typically trained in the various programming methodologies. Emphasis is too frequently placed upon the "craft" of programming rather than the processes in support of software engineering. IEEE software engineering standards can be used as supplemental training material for practitioners. These standards distill and present the consensus of software engineering best practices.

   In order for software process improvement to be successful, everyone involved should know how to effectively perform his or her roles. Many times, key individuals have not been effectively trained to perform to support the software engineering process. IEEE software and system engineering standards provide process and practice support to the software engineering knowledge as defined by the SWEBOK®. Each standard contains additional reference material that may be used in support of software engineering training activities. Many of the specific questions that practitioners have may be answered by turning to the IEEE Software Engineering Standards Collection. Table 4-1 provides a list of knowledge requirements and the IEEE software engineering standard that may be used to support the requirement.

Table 4-1. IEEE standards and training

Requirements	
How to document requirements	IEEE Std 830-1998, Recommended Practice for Software Requirements Specification
Test	
How to classify software anomalies	IEEE Std 1044-1993 (R2002), Standard Classification for Software Anomalies
How to select and apply software measures	IEEE Std 982.1-1988, Standard Dictionary of Measures to Produce Reliable Software
How to define a test unit	IEEE Std 1008-1987 (R2002), Standard for Software Unit Testing
What documentation is required in support of the testing process?	IEEE Std 829-1998, Standard for Software Test Documentation
Maintenance	
What maintenance activities are required prior to product delivery?	IEEE Std 1219, Standard for Software Maintenance
Communication	
How to communicate using consistent terminology	IEEE Std 610.12-1990 (R2002), Standard Glossary of Software Engineering Terminology
Configuration Management	
What describes the requirements and categories of information in support of configuration management planning?	IEEE Std 828-1998, Standard for Software Configuration Management Plans
Reviews and Audits	
Where to find information describing software audit procedures	IEEE Std 1028-1997 (2002), Standard for Software Reviews

CHARACTERIZE CURRENT AND DESIRED STATES

Characterizing the current and desired states for change is critical to the success and management of the change process. This provides the facts for laying out the implementation plan specifics and the degree of investments needed. Although the desired state is a management system that conforms to ISO 9001, alignment with business objectives is critical. This may result in retaining aspects of the current state.

The gap between current and desired state requires analysis by professionals knowledgeable in ISO 9001 for SE organizations, IEEE SE standards, and your SE methods, and the organization's culture for change management. The goal of this book is to provide your "gap analysis" team the insights into IEEE SE for ISO 9001; they should be allowed to attend ISO 9001 implementation training, which is typically two days long.

The desired states can vary. The authors caution against building a sterile, generic ISO 9001 management system structure and documentation for the purposes of passing the first external ISO 9001 audit in order to receive a certificate to please your customers. That approach normally results in a straitjacket system that requires continual manager's time to get SE personnel's conformance. SE staff traditionally views this type of artificial system as not useful for their activities and work products and actually getting in the way of doing real work. This book is geared toward creating a management system in which the SE personnel use their existing skills for SE products and services such as investigating, borrowing, designing, creating, transforming, building, revising, and collaborating. In addition, this management system can be understood by external stakeholders as it conforms to ISO 9001 requirements, follows ISO/IEC 90003, *Guidelines for the Application of ISO 9001:2000 to Computer Software,* aligned with IEEE 12207, *Software Life Cycle Processes,* and uses work products or artifacts aligned with IEEE SE standards. When using this book's approach for the desired state, future directions are possible to move up in organizational maturity by using the SEI CMMI model. Help is available from the book's authors who have published another book, *Practical Support for CMMI®-SW Software Project Documentation Using IEEE Software Engineering Standards.* This higher CMMI desired state is also adaptable for international customers inquiring about ISO/IEC 15504, *Process Assessments.*

Besides CMMI as a desired state, another desired state for information technology (IT) groups could be conformance to Control Objectives for Information and Related Technology (CobiT), which has 34 IT processes as well as 318 detailed control objectives and audit guidelines. CobiT provides a generally applicable and accepted standard for good IT security and control practices to support management's needs in determining and monitoring the appropriate level of IT security and control for their organizations.

Perform Gap Analysis

It is important to gauge how effectively process improvements have been implemented for continual process improvement to be successful. This can be determined through the development of a benchmarking appraisal to support gap analysis activities. This type of appraisal will provide a baseline for future process improvement efforts and will identify weaknesses and strengths.

Perform Self-Audit Using ISO 9001 Criteria

The components of ISO9001 can be used to form the basis for a gap analysis. These key components are described in detail in the ISO 9001:2000 standard. In this document, each required, expected, and informative component is addressed in detail. Supporting information for each requirement is also supplied in ISO/IEC 90003, Guide for Software. Use this detailed information when developing assessment matrixes. These matrixes can be used to identify areas of compliance or noncompli-

Table 4-2. Example of ISO 9001 Compliance Matrix

Req #	Source	Requirement	Satisfied by	Status (P/F/NI)
1	7.2.1.1	Customer-related requirements	Processes: development process, requirements elicitation, systems requirements analysis, software requirements analysis	Pass, Fail, or Needs Improvement (with comments)
2	7.2.1.2	Additional requirements determined by the organization	Processes: product quality software integrity levels, quality requirements and testing	

ance, or areas needing improvement. Table 4-2 provides sample data from a compliance matrix.

DEVELOP RECOMMENDATIONS

From the gap analysis outputs, the individual(s) performing the gap analysis should work with the Process Group to recommend actions that will move the organization from where it is to where you want it to be while identifying potential barriers to the effort. The results of the gap analysis, along with any additional recommendations, should be reported to the project stakeholders for reaffirmation.

For those projects having the context of customer or investor confidence that require only a simple realignment to ISO 9001 requirements with minor changes, the gap analysis sometimes surprises executives because of the degree of some gaps. At this point, it is appropriate to recognize any shift from customer or investor confidence to business improvement.

SET PRIORITIES

Following Steering Committee review and approval, the Process Group endorsed recommendations should be prioritized. The Process Group developed priorities of the recommended actions should be based on such things as:

- The availability of key resources
- Dependencies between the actions recommended
- Funding

Organizational realities influence priority setting. The Steering Committee has more control over the organizational realities than the Process Group, whose job is to surface these issues or caveats for Steering Committee support and decisions. Some common organizational realities are:

- Resources needed for change are limited
- Dependencies exist between recommended activities
- External factors may intervene
- The organization's strategic priorities must be honored

DEVELOP THE APPROACH

Once the Process Group has Steering Committee approval of the recommended actions, an approach should be developed that reflects both the established priorities and current realities. This approach should be a repeatable process and should consider the following:

- Strategy to achieve vision (set during initiating phase)
- Specifics of installing the new technology
- New skills and knowledge required of the people who will be using the technology, and recommended training
- Organizational culture
- Potential pilot projects
- Sponsorship levels
- Market forces

Goal-Driven Implementation

Any approach requires a strategy to be used by the organization to facilitate process institutionalization. This strategy should include targeted projects, key personnel, training, and, most importantly, schedules reflecting specific software process improvement targets. Goal-driven process improvement is the most effective. Identify both short- and long-term goals stating concise objectives and time periods and associate these goals with schedule milestones. Table 4-3 provides a sample time line for goal-driven process improvement.

PLAN ACTIONS

From the approach they have developed, the Process Group, they will next develop the action plans to implement the chosen approach. The action plans should describe the following:

- Deliverables, activities, and resources
- Decision points
- Milestones and schedule

- Risks and mitigation strategies
- Measures, tracking, and oversight

Many teams construct a project plan using programs such as Microsoft Project. This type of program maybe suitable for the first three bulleted items above, but will not be sufficient for the last two.

Baseline Processes

From the gap analysis outputs, the Process Group should define their process baseline, which is critical when implementing software processes that can be repeatable. Normally, baseline processes and descriptions can be managed in the Software

Table 4-3. Example goal-driven implementation time line

0–3 months
- Identify individuals responsible for software process improvement.
- Identify project managers who will be participating.
- Identify list of candidate projects.
- Solidify backing of senior management.
- Look at existing processes and make sure they are appropriate and reflect current business needs (small versus large projects) using ISO 9001 goals and IEEE software engineering standards.
- Define the formats for your process plans (Software Configuration Management Plan, Software Requirements Management Plan, Software Quality Assurance Plan) using IEEE software engineering standards and measure them against the ISO 9001 requirements.
- Get project members to provide feedback on process plans, review and incorporate feedback.
- Conduct a gap analysis against the ISO 9001 requirements.

3–6 months
- Create process document templates for project documentation based upon defined processes; projects will use these to develop their own plans (e.g., Software Development Plan, Software Requirements Specification).
- Conduct weekly/monthly status reports/reviews to gauge and report progress and provide areas for improvement.

6–9 months
- Conduct ISO9001-based reviews of the projects. It would be ideal to also include members from unselected projects to participate in these reviews, with reporting senior management.
- Provide feedback regarding project review, providing requirements for improvement to the projects.

9–12 months
- Conduct internal assessments with reporting to senior management.
- Provide feedback regarding project review, providing requirements for improvement to the projects.

Configuration Management (SCM) system as text documents. The SCM tool along with document creation and review responsibilities can meet the ISO 9001 Document Control requirement. Use IEEE standards to develop your baseline process documentation that addresses the ISO 9001 requirements. It may be helpful to define the initial process baseline at different levels of abstraction. One useful way of classifying these levels of abstraction is provided by the Basili [89] levels of abstraction:

- *Reference level*, with processes and responsibilities defined.
- *Contextual level*, describing connections among the processes to transfer information.
- *Implementation level*, combining the mapping of processes to parts of the organization.

Detailed information in support of this type of abstraction methodology and the supporting IEEE software engineering standards may be found in *Road Map to Software Engineering—A Standards-Based Guide* [89]. This additional information in support of process abstraction can be useful when trying to define a process baseline.

It is also important to evaluate and identify any potential tools that may be used in support of process automation, keeping in mind that though a tool is not a substitute for a process, it can facilitate its implementation. An ideal candidate area for this type of automation is Software Configuration Management (SCM). There are a number of widely marketed tools that support the documentation and control of various types of software configuration items.

Once a process baseline has been established, an action plan should be formulated. The action plan provides a map of the path forward—improvement of the baseline processes. Table 4-4 provides a sample from an action plan.

Take advantage of the information provided by the IEEE Software Engineering Standards Collection. Many of these standards provide documentation templates and describe in detail what individual project support processes should contain. Think of the standards as an in-house software process consultant who has recom-

Table 4-4. Example action plan

Process area	Weakness or area for improvement	Short description of how to address	Project point of contact	Due or resolution date	Resources/ support required	Risks/ contingency
Requirements Management	7.2.1.1. The establishment of a method for traceability of the requirements to the final product	Create requirements traceability matrix in SCM with links to and from reviews	Jim Smith	4Q2006	SCM Req. doc., Req. Rev., Client Rev.	Review time and rework.

mended, based upon years of experience, the proper methodologies and techniques to be used in support of software development.

CREATE SOLUTION

The Solution arising from the approved gap analysis recommendations will result in defining and implementing new and/or revised procedures, templates, forms, and checklists, along with measurable objectives that support the organization's quality objectives.

The Process Group will bring together everything available to create a "best guess" solution specific to organizational needs, for example:

- Existing tools, processes, knowledge, skills, and so on
- New knowledge, information, and so on
- Outside help

The solution should:

- Identify performance objectives
- Finalize the pilot/test group
- Construct the solution material and train with the pilot/test group
- Develop draft plans for pilot/test and implementation
- Technical working groups develop the solution with pilot/test group

The SE management system conforming to ISO 9001 will include a hierarchy of documents and records necessary to run the SE business. An organization collection of policies can be the heart of the ISO 9001 Quality Manual. Also included are the key business processes. Separately linked are the implementation of policies and key business processes into required procedures and records. SE processes create work products or artifacts that normally require team member and/or customer review and acceptance. Artifact templates, checklists, and record forms can facilitate these reviews and acceptance. In addition to policies, procedures, artifacts, and records are the process measures. Objective measures can cover time interval, investments, costs, output quantities, and rework indicators such as defect counts and feedback.

PILOT/TEST SOLUTION

The pilot/test group puts the solution in place on a trial basis to learn what works and what does not. The length of the pilot should include at least one process cycle to exercise inputs, outputs, controls, and measurements. These pilot steps include:

- Train the pilot or test group.
- Perform the pilot or test.

- Gather feedback.
- Perform Process Group audit according to applicable desired state requirements.
- Review Process Group review findings and feedback.

REFINE SOLUTION

The Process Group will revisit and modify the solution to incorporate new knowledge and understanding. Iterations of pilot/test and refinement may be necessary before arriving at a solution that is deemed satisfactory to meet the expected process measurement goals.

IMPLEMENT SOLUTION

When the solution is deemed workable and sufficient, implement it throughout the organization:

- Adjust the plan as necessary.
- Execute the plan.
- Conduct post implementation analysis.
- Various approaches may be used: top-down, lateral, or staged.

ANALYZE AND VALIDATE

The Process Group should audit the organization according to applicable desired state requirements. Next, the Process Group determines what was actually accomplished by the improvement effort:

- In what ways did it or did it not accomplish its intended purpose?
- What worked well?
- What could be done more effectively or efficiently?

Validation includes comparing what was accomplished with the intended purpose for undertaking the change. The Process Group should report back to the Steering Committee on the summarization of lessons learned regarding processes used to implement the IDEAL process. The Steering Committee should take part in the organization's celebration.

ISO 9001 Registration Steps

If the conformance to ISO 9001 requires an external registration, the following steps should be taken:

- ISO 9000 awareness program
- Management review of audit findings and proposed corrective and preventable actions
- Implementation of management review actions
- Preaudit visit by registrar
- Registrar audit
- Corrective actions to registrar audit findings
- Registrar approves of corrective actions
- Registrar approves ISO 9001 registration

PROPOSE FUTURE ACTIONS

The Process Group should review, catalog, and save reusable intellectual assets in the SCM for future organizational process assets. The Process Group should develop recommendations concerning management of future change efforts using IDEAL:

- Improving an organization's ability to use IDEAL
- Address a different aspect of the organization's business

IMPLEMENTATION PITFALLS

The implementation of ISO 9001 is fraught with some common pitfalls. It is important to remember that this standard is high level and not prescriptive. Too often, organizations implement elaborate process control and improvement, spending considerable time and effort, without realizing significant improvements in product cost, quality, or cycle time. In order for organizations to fully benefit from process improvement activities, time must be spent focusing on the definition of current practices and procedures. Only then, after this initial process definition is complete, should organizations move forward to begin to improve their existing processes.

Being Overly Prescriptive

Implementing the ISO 9001 requirements verbatim can be costly and may not reflect the specific process or business requirements of an organization. This approach can result in an overly prescriptive process that will increase cost and slow down product cycles. This can rapidly destroy the credibility of the process improvement implementation with management and the software development staff. Carefully analyze what areas of process improvement will provide the most positive impact on existing programs; implement these first and management will see the tangible results and continue their support. Small projects may require less formality in planning than large projects, but all components of the standard should be addressed by every software project. Components may be included in the project-level documentation, or they may be merged into a system-level or business-level plan, depending upon the complexity of the project.

Remaining Confined to a Specific Stage

Be careful to not examine the implementation of each ISO 9001 requirement in isolation. Many organizations that are just beginning a process improvement effort often delay implementation of a risk assessment program. Risk avoidance can be a critical factor during software development and an organization can significantly benefit if this area is addressed early on.

Also, be aware that all projects within an organization may be at different places in their implementation of ISO 9001 requirements. Most organizations identify several pilot projects and move them through their development life cycles. The experience gained from this type of implementation is then carried throughout the entire organization to the remaining efforts.

Documentation, Documentation

Be careful not to generate policies and procedures to simply satisfy ISO 9001 requirements. Generating documentation for the sake of documentation is a waste of time and resources. Policies and procedures should be developed at the organizational level and used by all projects, only requiring documentation where there is deviation from the standard. The pairing of IEEE standards, which are prescriptive and describe software engineering minimums, with these models can reduce this risk significantly.

Lack of Incentives

ISO 9001 references the Management Representative, whom many companies call the Quality Manager/Director. This book's approach uses the Steering Committee with technical representatives and the Process Group (PG) of technical management. Many times, the success of the process improvement initiatives is placed on the shoulders of these groups. The PG should act as mentor, guiding the software projects through continual process improvement. Each selected project should be required to follow the recommendations of these groups. Process improvement is most effective and the results more permanent when the projects are stakeholders in the software improvement process.

No Measurements

If your organization has effectively implemented process improvement, you can realize improved business results. However, too often organizations implement processes in support of ISO 9001 requirements just to meet the process goals for certification purposes. Inefficient implementation will be the result when organizational needs are not made a priority. Process performance can be static, or even degrade, relative to business goals. The standard does not require substantial process performance measurements. Without a solid understanding of the cost–benefit ratio and its relationship to business results, it can be easy to lose corporate support.

CONCLUSION

Process improvement can be intimidating. Many times, the task of process improvement comes in the form of a directive from senior management, or as a customer requirement, leaving those assigned with a feeling of helplessness. However, all those practicing as software engineers should desire to evolve out of the chaotic activities associated with uncontrolled software processes, and the associated required heroic efforts. When software engineering processes are under basic management control and there is an established management discipline, then this provides benefits to all involved. When ISO 9001 is used in conjunction with IEEE software engineering standards, the customer may be assured of a lower risk of fail-

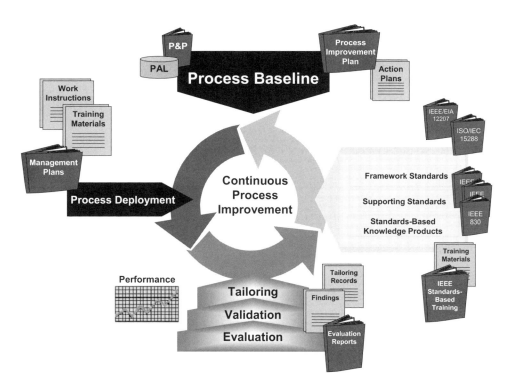

Figure 4-2. Standards support continual process improvement [67].

ure, the organization is provided with accurate insight into the effort, management can more effectively identify and elevate development issues, and team members can work to efficiently managed baselines.

The requirements of ISO 9001 are broad. No single IEEE standard can be used in isolation to support these requirements. Rather, a subset of the available IEEE Software Engineering Standards should be employed in combination (refer to Figure 4-2) to provide effective support for ISO 9001 activities. IEEE software engineering standards can be used to provide detailed, prescriptive, support for ISO 9001 process definition and improvement activities.

Chapter **5**

12207 Primary Life Cycle Processes and ISO 9001

SOFTWARE LIFE CYCLE (SLC) SELECTION AND DESIGN

Software engineering (SE) organizations conforming to ISO 9001 start with Clause 4.1, General Requirements, which states that the quality management system must be established, documented, implemented, and maintained; its effectiveness must be continually improved in accordance with the requirements of ISO 9001. To implement the system, an organization must:

- Identify processes needed for the quality management system (and their application throughout the organization).
- Determine process sequence and interaction.
- Determine criteria and methods for process operation and control.
- Ensure that resources and supporting information are available.
- Monitor, measure, and analyze these processes.
- Implement actions to achieve planned results and continual process improvement.

Furthermore, the ISO standard notes that quality management system processes should include processes for management activities, provision of resources, product realization, and measurement. The scope of "product realization" is identical to SE. Thus, this ISO 9001 requirement directs the SE organization to define its software life cycle (SLC) models. The SLC starts at its inception continues to its completion, and ends with its withdrawal or replacement. Software engineering practitioners are familiar with the many SE processes and their relationships during the SLC. Software product realization planning is covered in ISO 9001, Clause 7.1, Planning of Product Realization.

The processes needed for product realization must be planned and developed by the organization. The planning must be consistent with other requirements of the quality management system and be documented in a suitable form for the organization. The planning must also determine, as appropriate, the:

- Quality objectives and product requirements
- Need for processes, documents, and resources
- Verification, validation, monitoring, inspection, and test activities
- Criteria for product acceptance
- Records as evidence that the processes and resulting product meet requirements

Implementation of this clause is where processes, activities, and tasks should be planned and performed using software life cycle models suitable to the nature of a software project, considering size, complexity, safety, risk, and integrity. IEEE 12207.0, Software Life Cycle Processes was developed to describe the framework of the comprehensive set of software life cycle processes. The related IEEE 1074, Developing a Software Project Life Cycle Process, describes the method of selecting, implementing and monitoring a SLC for a SE project. There are several types of SLCs, such as waterfall, modified waterfall, V-shaped, incremental, spiral, synchronize and stabilize, rapid prototype, and code and fix.

Waterfall

The waterfall model uses a sequence of development stages in which the output of each stage becomes the input for the next. The focus is on control. This is most common with maintenance releases, in which the control of changes is important to the customer.

Modified Waterfall

The modified waterfall or fountain model recognizes that although some activities cannot start before others—such as you need a design before you can start coding—there is a considerable overlap of activities throughout the development cycle.

V-Shaped

The V-shaped model is just like the waterfall model. The V-shaped life cycle is a sequential path of execution of processes. Each phase must be completed before the next phase begins. Testing is emphasized in this model more so than the waterfall model, though. The testing plans, procedures, and data are developed early in the life cycle before any coding is done, during each of the phases preceding implementation.

Incremental

The incremental model is an intuitive approach to the waterfall model. The incremental model divides the product into multiple builds, in which sections of the project are created and tested separately. Multiple development cycles take place here, making the life cycle a "multiwaterfall" cycle. Cycles are divided up into smaller, more easily managed iterations. Each iteration passes through the requirements, design, implementation, and testing phases. This approach will likely find errors in user requirements quickly, since user feedback is solicited for each stage and because code is tested sooner after it is written.

Spiral

The spiral life-cycle model is the combination of the classic waterfall model and an element called risk analysis. This model is very appropriate for new projects and large software projects. It consists of four main parts or blocks—Planning, Risk Analysis, Engineering, and Customer Evaluation—and the process is shown by a continuous spiral loop going from the outside toward the inside. The baseline spiral starts in the planning phase, where requirements are gathered and risk is assessed. Each subsequent spiral builds on the baseline spiral. The spiral model emphasizes the need to go back and reiterate earlier stages a number of times as the project progresses. Each cycle produces an early prototype representing a part of the entire project. This approach helps demonstrate a proof of concept early in the cycle, and it more accurately reflects the disorderly, even chaotic, evolution of technology.

Synchronize and Stabilize

The synchronize and stabilize method combines the advantages of the spiral model with technology for overseeing and managing source code. This method allows many teams to work efficiently in parallel.

Rapid Prototype for New Projects

In the rapid prototyping (sometimes called rapid application development) model, initial emphasis is on creating a prototype that looks and acts like the desired product in order to test its usefulness. The prototype is an essential part of the requirements determination phase, and may be created using tools different from those used for the final product. Once the prototype is approved, it is discarded and the "real" software is written.

Code-and-Fix

Code-and-Fix is the crudest of the methods. If you do not use a methodology, it is likely that you are doing code-and-fix. Write some code and then keep modifying it until the customer is happy or the delivery date arrives. Without planning, this is very open ended and can by risky. Code-and-fix rarely produces useful results. It is very dangerous, as there is no way to assess progress, quality, or risk.

IEEE 12207 PROCESSES

IEEE 12207 standards establish a framework for the life cycle of software. The life cycle begins with an idea or need that can be satisfied wholly or partly by software and ends with the retirement of that software. The framework architecture is built with a set of processes. Interrelationships exist among these processes and each process is placed under the responsibility of an organization or a party in the software life cycle. An organization may employ an organizational process to establish, control, and improve a life cycle process. Figure 5-1 provides an overview of these processes.

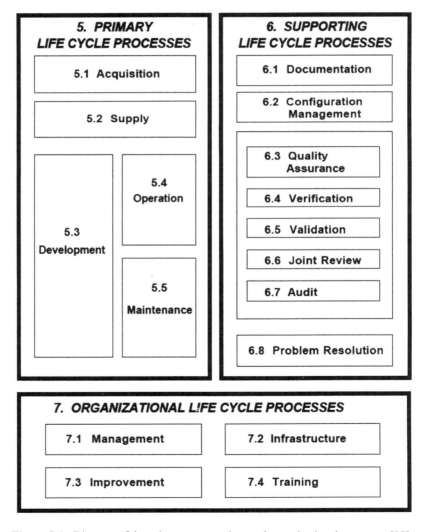

Figure 5-1. Diagram of the primary, supporting, and organizational processes [37].

The IEEE 12207 standards group the activities that may be performed during the life cycle of software are divided into five primary processes, eight supporting processes, and four organizational processes. Each life cycle process is divided into a set of activities; each activity is further divided into a set of tasks. Figure 5-2 describes these primary, supporting, and organizational SLC processes.

The five primary processes serve primary parties during the life cycle of the software. A primary party is one that initiates or performs the development, operation, or maintenance of software products. These primary parties are the acquirer, the supplier, the developer, the operator, and the maintainer of software products. These primary processes relate to ISO 9001 Section 7, Product Realization.

The eight supporting life cycle processes support other processes as an integral part with a distinct purpose and contribute to the success and quality of the software project. A supporting process is employed and executed, as needed, by another process. These supporting processes are covered throughout ISO 9001. Chapter 6 covers these processes.

The four organizational life cycle processes are employed by an organization to establish and implement an underlying structure made up of associated life cycle processes and personnel and continuously improve the structure and processes. They are typically employed outside the realm of specific projects and contracts; however, lessons from such projects and contracts contribute to the improvement of the organization. These organizational processes are directly related to ISO 9001, Section 5, Management Responsibility; Section 6, Resource Management; and Section 8, Measurement, Analysis and Improvement. Chapter 7 covers these processes.

The five 12207 primary processes, shown in Table 5-1, are performed by distinct parties: the acquirer, the supplier, the developer, the operator, and the maintainer of software products.

ACQUISITION

The acquisition process defines the activities of the acquirer. The process begins with the definition of the need to acquire a system, software product or software service. The process continues with the preparation and issue of a request for proposal, selection of a supplier, and management of the acquisition process through to the acceptance of the system, software product, or software service. Table 5-2 provides a list of the acquisition process objectives.

ISO 9001 Goals

ISO 9001 describes requirements in support of the acquisition of both software products and services. In this ISO standard, Clause 7.4, Purchasing, describes the processes of supporting information and verification activities required in support of product or services acquisition.

ISO 9001, Clause 7.4.1, Purchasing Process, requires the organization to ensure that the purchased software product conforms to specified purchase requirements.

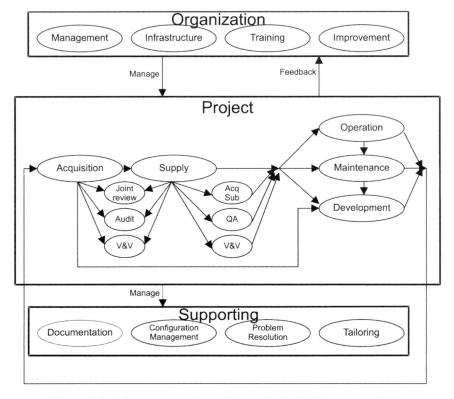

Figure 5-2. Primary, supporting, and organizational SLC processes [37].

Table 5-1. The five 12207 primary processes [37]

Acquisition process
Defines the activities of the acquirer, the organization that acquires a system, software
product or software service.

Supply process
Defines the activities of the supplier, the organization that provides the system, software
product or software service to the acquirer.

Development process
Defines the activities of the developer, the organization that defines and develops the software product.

Operation process
Defines the activities of the operator, the organization that provides the service of operating a computer system in its live environment for its users.

Maintenance process
Defines the activities of the maintainer, the organization that provides the service of
maintaining the software product; that is, managing modifications to the software product
to keep it current and in operational fitness. This process includes the migration and retirement of the software product.

Table 5-2. Acquisition process objectives

a) Develop a contract, including tailoring of the standard that clearly expresses the expectation, responsibilities, and liabilities of both the acquirer and the supplier.

b) Obtain products and/or services that satisfy the customer need.

c) Manage the acquisition so that specified constraints (e.g., cost, schedule, and quality) and goals (e.g., degree of software reuse) are met.

d) Establish a statement of work to be performed under contract.

e) Qualify potential suppliers through an assessment of their capability to perform the required software.

f) Select qualified suppliers to perform defined portions of the contract.

g) Establish and manage commitments to and from the supplier.

h) Regularly exchange progress information with the supplier.

i) Assess compliance of the supplier against the agreed-upon plans, standards, and procedures.

j) Assess the quality of the supplier's delivered products and services.

k) Establish and execute acceptance strategy and conditions (criteria) for the software product or service being acquired.

l) Establish a means by which the acquirer will assume responsibility for the acquired software product or service.

The type and extent of control applied to the supplier and purchased product depends upon the effect of the product on the subsequent realization processes or the final product. For commercial-off-the-shelf (COTS) software, almost no control is applied to the supplier. On the other hand, for custom software development, the acquirer takes on the role of the knowledgeable customer, deeply involved with the supplier throughout their SLC. This involvement includes participation as shown in ISO 9001, Clause 7.2.2, Review of Requirements Related to the Product. For either extreme, COTS or custom software development, this process requires evaluation and selection of the supplier based on established criteria and a record kept on the evaluation and actions taken. IEEE 1062, Recommended Practice for Software Acquisition, defines the Software Acquisition Plan (SAP). Objective evidence of supplier selection can be documented in a Decision Matrix.

ISO 9001 uses Clause 7.4.2, Purchasing Information, to require the organization to describe the software product. Beyond COTS purchases, this information may contain product and process approval requirements, operational personnel requirements, and supplier quality management system requirements. These requirements must be communicated to the supplier in a request for proposal (RFP) and should be adequate for a proper response. IEEE 1062, the Recommended Practice for Software Acquisition provides the requirements for the contents of an RFP. For more extensive automation of operations, IEEE 1362, Concept of Operations (ConOps), defines the current system or situation, justification for and nature of changes, concepts for the proposed system, operational scenarios, summary of impacts, and analysis of the proposed system.

ISO 9001, Clause 7.4.3, Verification of Purchased Product, requires the organization to test their software acquisitions to insure, that they meet the specified purchase requirements. IEEE 829, Software Test Documentation, can be used for customer acceptance test planning.

IEEE standards support the acquisition process requirements very directly; the project documentation in support of supply and acquisition activities described in this section is derived from IEEE Std 1062, Software Acquisition Plan.

Software Acquisition Plan

A clear description of how adequately IEEE Std 1062 and IEEE Std 12207 support the ISO 9001 products and services acquisition requirements can be seen in a description of the *Recommendations for Software Acquisition,* Appendix C. IEEE Std 12207 and IEEE Std 1062 support each step with prescriptive detail. The modification of the recommended Software Acquisition Plan (SAP) table of contents to support the goals of ISO 9001 more directly is shown in Table 5-3. Appendix C, Software Process Work Products, provides additional information relating to the work products associated with software acquisition activities. These include a discussion of the nine steps of acquisition, an organizational acquisition strategy checklist, supplier evaluation criteria, and supplier performance standards.

Software Acquisition Plan Document Guidance

The following provides section-by-section guidance in support of the creation of a SAP. The SAP should be considered to be a living document and should change to reflect any process improvement activity. This guidance should be used to help define a software acquisition process and should reflect the actual processes and procedures of the implementing organization. Additional information is provided in the document template, *Software Acquisition Plan.doc,* which is located on the CD-ROM included in this book.

> *Introduction.* The introduction to the SAP should describe the specific purpose, goals, and scope of the software acquisition effort. All associated requirements and plans should be identified. The type of contract and support concept should be identified.
>
> *References.* This section should include all documents all supporting documents supplementing or implementing the SAP, including other plans, processes, or task descriptions that elaborate details of this plan.
>
> *Definitions and Acronyms.* This section of the SAP should define or reference all terms unique to the development and understanding of the SAP. All abbreviations and notations used in the SAP should also be described.
>
> *Software Acquisition Overview.* This section should describe the acquiring organization, acquisition schedule, and associated resources. All responsibilities, tools, techniques, and methods necessary to perform the software acquisition process should also be described.

Table 5-3. Software acquisition plan document outline

Title Page
Revision Page
Table of Contents
1. Introduction
2. References
3. Definitions and Acronyms
4. Software Acquisition Overview
4.1 Organization
4.2 Schedule
4.3 Resource Summary
4.4 Responsibilities
4.5 Tools, Techniques, and Methods
5. Software Acquisition Process
5.1 Planning Organizational Strategy
5.2 Implementing the Organization's Process
5.3 Determining the Software Requirements
5.4 Identifying Potential Suppliers
5.5 Preparing Contract Documents
5.6 Evaluating Proposals and Selecting the Suppliers
5.7 Managing Supplier Performance
5.8 Accepting the Software
5.9 Using the Software
6. Software Acquisition Reporting Requirements
7. Software Acquisition Management Requirements
7.1 Anomaly Resolution and Reporting
7.2 Deviation Policy
7.3 Control Procedures
7.4 Standards, Practices, and Conventions
7.5 Performance Tracking
7.6 Quality Control of Plan
8. Software Acquisition Documentation Requirements

Organization. This subsection of the SAP should describe the organization of the acquisition effort. This description of the organization should include the approval hierarchy and points of contact.

Schedule. This subsection should describe the schedule of the acquisition to include all milestones. If this schedule is included as part of the acquiring organization's software project management plan, this plan may be referenced.

Resource Summary. This subsection should describe all staffing, facilities, tools, finances, and any special procedural requirements that are needed in support of the software acquisition effort.

Responsibilities. This subsection should provide an overview of acquisition responsibilities.

Tools, Techniques, and Methods. This subsection should describe all documentation, tools, techniques, methods, and environments to be used during the acquisition process. Acquisition, training, support, and qualification information for each should be provided. The SAP should document the measures to be used by the acquisition process and should describe how these measures support the acquisition process.

Software Acquisition Process. This section should identify all actions to be performed for each of the software acquisition steps shown:

1. Planning organizational strategy
2. Implementing the organization's process
3. Determining the software requirements
4. Identifying potential suppliers
5. Preparing contract documents
6. Evaluating proposals and selecting the suppliers
7. Managing supplier performance
8. Accepting the software
9. Using the software

For further descriptions, refer to the table in Appendix C of this book entitled "Recommendations for Software Acquisition." Any additional information required in support of the software acquisition process should be included as deemed necessary.

Software Acquisition Reporting Requirements. This section should describe all reporting requirements. This should include a description of the reporting in support of acquisition status and risk.

Software Acquisition Management Requirements. This section should describe all procedures and processes in support of anomaly resolution and reporting. References should be provided to other plans describing the management of the software acquisition process. Subsections to this section are:

Anomaly Resolution and Reporting. This subsection describes the method of reporting and resolving anomalies, including all reporting and resolution criteria.

Deviation Policy. This subsection should describe all procedures and forms used if deviation is required. All deviation approval authorities should also be identified.

Control Procedures. This subsection should describe all procedures and processes supporting the configuration, protection, and storage of associated software products.

Standards, Practices, and Conventions. This subsection of the plan should describe all standards, practices, and conventions used in the development of this plan or in support of the acquisition process.

Performance Tracking. This subsection should describe the processes supporting performance monitoring. All items tracked should be identified.

All reporting procedures should also be described, including reporting format.

Quality Control of the Plan. This subsection should describe the processes supporting the development and maintenance of this plan. Other project plans may be referenced in support of this section (e.g., Software Quality Assurance Plan).

Software Acquisition Documentation Requirements. All documentation requirements required in support of the acquisition process should be described here. An appendix providing examples of all required document formats could be included.

See the Appendix C for templates useful in selecting the best suppliers:

Make/Buy Decision Matrix
Alternative Solution Screening Criteria Matrix
Cost–Benefit Ratio

Concept of Operations

The following information is based on IEEE Std 1362, IEEE Guide for Information Technology—System Definition—Concept of Operations (ConOps) Document. The following provides section-by-section guidance in support of the creation of a ConOps document. This guidance should be used to help define a requirements management process and should reflect the actual processes and procedures of the implementing organization. IEEE Std 12207.0 provides information in support of the development of a concept of operations description. The information provided in this section is in conformance with IEEE Std 12207.

Additional information is provided in the document template, *ConOpsDocument.doc,* which is located on the companion CD-ROM. Table 5-4 provides an example of a suggested document outline supporting ISO 9001 requirements.

Concept of Operations (ConOps) Document Guidance

A ConOps document is used to represent the viewpoint of the system, or software, user. The ConOps should effectively describe the system characteristics to all participants and how these characteristics meet the mission and objectives of the organization. It includes the following:

Document Overview. A description of the document, the target audience, and the rationale for its use should be presented in this section.

System Overview. This section of the document should provide an overview of the system and the intended audience and purpose. A diagram of the system providing an overview of the functionality is often useful when trying to convey this type of information.

Table 5-4. Concept of operations (ConOps)
document outline

Title page
Revision Page
Table of Contents
1. Scope
    1.1   Document Overview
    1.2   System Overview
2. Referenced Documents
3. Definitions and Acronyms
4. Operating Procedures
    4.1   Background
    4.2   Operational Policies and Constraints
    4.3   Current Operating Procedures
    4.4   Modes of Operation
    4.5   User Classes
    4.6   Support environment
5. Change Justification
    5.1   Changes Considered
    5.2   Assumptions and Constraints
6. Proposed System
7. Operational scenarios
8. Impact Summary
    8.1   Operational Impact
    8.2   Organizational Impact
9. System Analysis
    9.1   Summary of Improvements
    9.2   Disadvantages and Limitations
    9.3   Alternatives
Appendices

Referenced Documents. This section should provide all supporting documentation used in the development of this document. All documentation referenced by the ConOps should also be listed in this section.

Definitions and Acronyms. This section should provide a list of all definitions and acronyms unique to this document and critical to understanding its content.

Operating Procedures. This section should describe the current operating procedures. These procedures can be based upon an existing system or may describe manual procedures. If there are no current operating procedures, then the ideal should be described. Subsections to this section are:

Background. This subsection should provide the reader with information in support of the project background, objectives, and scope. The information provided here should describe the problem set and proposed solution.

Operational Policies and Constraints. This subsection should provide information in support of all operational policies and constraints as they apply to the current system or situation. As defined by IEEE Standard 1362, IEEE Guide for Information Technology—System Definition—Concept of Operations (ConOps) Document, "Policies limit decision-making freedom but do allow for some discretion. Operational constraints are limitations placed on the operations of the current system. Examples of operational constraints include the following:

- A constraint on the hours of operation of the system, perhaps limited by access to secure terminals.
- A constraint on the number of personnel available to operate the system.
- A constraint on the computer hardware (for example, must operate on computer X).
- A constraint on the operational facilities, such as "office space."

Current Operating Procedures. This subsection should provide a detailed description of the current system or situation. This should include a description of the operating environment, major system components, required interfaces, performance requirements, and desired features. Presenting this information to the system user in graphical format is useful when attempting to describe a system or situation. It may be useful to include items such as schedules, charts, functional and/or data flow, or workflow diagrams.

This section should also provide a description of all operational requirements in language that the system user would understand. All required facilities, material, hardware and software, and personnel should be described.

Modes of Operation. This subsection should provide detailed information describing all required modes of operation for the system or situation.

User Classes. This subsection should provide a description of all user classes. This should include a description of the organizational structure, user profiles with associated roles and required skill sets, and required interaction between user classes. This subsection should also describe all key personnel associated with the project. Behavior diagrams can include the use case diagram (used by some methodologies during requirements gathering), sequence diagram, activity diagram, collaboration diagram, and statechart diagram.

Support Environment. As applicable, this subsection should describe all support activity required for the maintenance of the system or situation.

Change Justification. This section should provide a description of all deficiencies of the current system or situation. This section should provide a description of all identified problems in terms that can be easily understood by the user. This section should provide a description of all known issues, proposed changes, and a justification for the proposed change. It is helpful to also provide an Appendix to this document that lists all proposed changes in order of importance. A common method used is to categorize items as essential (must have), desired (nice to have), or optional. This ranking is helpful during the translation the items described in the ConOps document to a more formal requirements specification. Subsections to this section are:

Changes Considered. This subsection should provide information regarding all features (changes) considered but not included in the proposed software application. The impact of not including these features should be described along with any plans for future adoption.

Assumptions and Constraints. This subsection should address all assumptions and constraints that will affect users during development and operation of software application. It is important to tie assumptions to assessments of impact. Describe what the new system will provide in terms of performance gains, interfaces to external systems, or schedule impact.

Proposed System. This section should describe the proposed system providing a high-level, broad system description. This description should be solution oriented. It should focus on the user needs and how the proposed system will support the needs of the user community. Any description of the proposed system should address the characteristics of the operating environment, the performance characteristics, all interface requirements, the capabilities and functions, a description of the dataflow requirements, associated cost and risk, and quality requirements.

It is important to keep in mind that the ConOps should be written using common language. Avoid computer-related jargon and use graphics where possible.

Operational Scenarios. This section should provide a description, or series of descriptions, of how the proposed system will operate. Scenarios may also be used to describe what the system will not do. Scenarios should be used to help readers understand how system functionality will support operational requirements.

Impact Summary. This section should describe the positive and negative perceived impacts of the proposed system. All predeployment, deployment, and training activities should be described with any accompanying impact description. Providing this type of information will help organizations prepare for any disruptions encountered during system deployment. Subsections to this section are:

Operational Impact. This subsection should describe all anticipated impacts to the system user during system operation. System deployment may require a change in current policy or procedure, and these should be addressed.

Organizational impacts. This subsection should describe all anticipated organizational impacts. All impacts associated with users, system development, and system deployment during the operation of the proposed system should be addressed.

System Analysis. This section should provide an analysis of the proposed system. A summary of all improvements, disadvantages and limitations, and alternatives considered as relating to the proposed system should be described. Subsections to this section are:

Summary of Improvements. This subsection should provide an evaluation of all improvements to existing processes, or practices, to be provided by the proposed system. This summary should include a description of any new or enhanced capabilities.

Disadvantages and Limitations. This subsection should describe any perceived disadvantages or limitations presented by the deployment of the proposed system.

Alternatives. This subsection should present a description of any alternatives considered. All alternatives considered, but rejected, should be documented with a rationale for nonacceptance.

Appendices. To facilitate ease of use and maintenance of the ConOps document, some information may be placed in appendices to the document. Each appendix should be referenced in the main body of the document where that information would normally have been provided.

Decision Tree Analysis

Decision analysis tools and techniques may be used to support the technical decisions that must be made during the life cycle of a software project. Some of these decisions include the type of architecture, determining whether to build or buy, product design, platform type, life cycle selection, and testing approaches. Table 5-5 provides the steps typical to the decision process and associated sample questions.

Decision trees are often useful when attempting to choose between several courses of action. They provide a highly effective structure for the evaluation of options, associated outcomes, and associated risk and benefit. A decision tree begins with a decision that is represented by a small square toward the left of the drawing area. Lines are drawn out to the right from this initial decision box, each line representing an alternative solution.

Consider the results of each decision path. If the alternative presented results in uncertainty, draw a circle noting the uncertainty above the circle. In a decision tree, squares represent decision and circles represent uncertain outcomes. If the alternative results in another decision, draw another square, noting the decision above the square. This process should be repeated until all possibilities are exhausted.

The decision tree must now be evaluated. Begin by identifying a percentage value for each of the possible outcomes. Also assign a dollar amount or value to each alternative outcome. Next, perform the calculations associated with each node of the decision tree and record the result. When calculations are based upon uncertainties (circled items) multiply the value of the outcome by the probability shown. After the estimated outcome, or benefit, is calculated it is important to then subtract the estimate for all associated costs. The decision tree will then re-

Table 5-5. Basic decision analysis and resolution process [90]

Step in process	Sample questions
1. Draft Decision Statement	In what situation do we need to take action? What are we trying to achieve?
2. Establish Decision Objectives	What are the anticipated results? What resources do we have to work with?
3. Objectives: Required or Desired?	What is critical to the success of the decision and can this be measured?
4. Value the Desired Objectives	What is the value of each objective? What is the value scale?
5. Develop the Alternatives	Are there possible alternative solutions?
6. Test Alternatives against Required Objectives	How do the alternatives compare against the objectives?
7. Score Alternatives against Desired Objectives	What is the value of each alternative? What is the value scale?
8. Determine Risks	What are the associated probabilities and impact of the risk?
9. Select Best Alternative	What alternative provides the most benefit for the least risk?

flect net benefit values, which may be used in the decision making process. An example of a basic decision tree is presented in Figure 5-3. An example of the recorded outcomes, or benefit, for a new product through development is provided in Table 5-6.

Decision trees can help provide objective insight when analyzing information for decision making. They help to clarify the problem by laying out all the options. All probabilities are factored into the various alternatives. The key to an effective decision-making tool is that they help individuals make objective decisions.

SUPPLY

The supply process contains the activities and tasks of the supplier. The process may be initiated either by a decision to prepare a proposal to answer an acquirer's request for proposal or by signing and entering into a contract with the acquirer to provide the system, software product, or software service. The process continues with the determination of procedures and resources needed to manage and assure the success of the project, including development of project plans and execution of the plans through delivery of the system, software product, or software service to the acquirer. Table 5-7 provides a list of the supply process objectives.

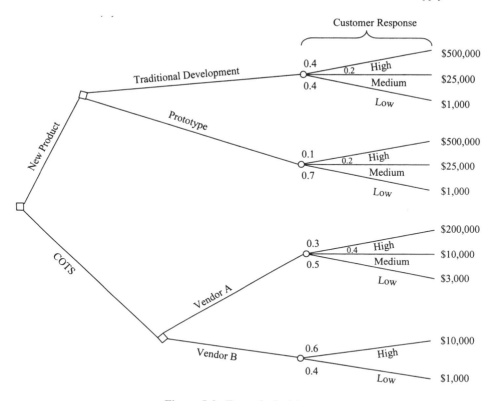

Figure 5-3. Example decision tree

Table 5-6. Example decision tree results

Node	Benefit level	Benefit calculations	Benefit subtotal	Outcome total	Estimated cost	Net benefit
Traditional development	High Medium Low	0.4 × 500,000 0.4 × 25,000 0.2 × 1,000	200,000 10,000 200	210,000	75,000	135,000
Prototype	High Medium Low	0.1 × 500,000 0.2 × 25,000 0.7 × 1,000	50,000 5,000 700	55,700	40,000	15,700
Vendor A	High Medium Low	0.3 × 200,000 0.4 × 10,000 0.5 × 3,000	60,000 4,000 1,500	65,500	15,000	50,000
Vendor B	High Low	0.6 × 10,000 0.4 × 1,000	6,000 400	6,400	0	6,400

Table 5-7. Supply process objectives

---

a) Establish clear and ongoing communication with the customer.

b) Define documented and agreed-upon customer requirements, with managed changes.

c) Establish a mechanism for ongoing monitoring of customer needs.

d) Establish a mechanism for ensuring that customers can easily determine the status and disposition of their requests.

e) Determine requirements for replication, distribution, installation, and testing of the system containing the software or stand-alone software product.

f) Package the system containing software or the stand-alone software product in a way that facilitates its efficient and effective replication, distribution, installation, testing, and operation.

g) Deliver a quality system containing software or stand-alone software product to the customer, as defined by the requirements, and install in accordance with the identified requirements.

ISO 9001 Goals

ISO 9001 describes requirements in support of the supplier of both software products and services. The first three clauses involve working with the customer. Clause 7.2.1, Determination of Requirements Related to the Product, requires determining customer requirements, both specified for the product (including delivery and post-delivery activities) and not specified for the product (but needed for specified or intended use, where known). Furthermore the supplier must also determine any statutory and regulatory requirements related to the product. The specific ISO 9001 Clause 7.2.2, Review of Requirements Related to the Product, directs the organization to review the software product requirements before committing to supply the software product to the customer in order to:

- Ensure that software product requirements are defined
- Resolve any requirements differing from those previously expressed
- Ensure its ability to meet the requirements [47]

The results of the review, and any subsequent follow-up actions, must be maintained. When the software requirements are not documented, they must be confirmed before acceptance. If product requirements are changed, the organization must ensure that relevant documents are amended and relevant personnel are made aware of the changed requirements. A common outcome of this process is the completed request for proposal (RFP). IEEE Standard 1062, Software Acquisition, provides further details on the RFP and its role in the acquisition process.

Another supplier responsibility from ISO 9001 Clause 7.2.3, Customer Communication, is that the organization must determine and implement effective arrangements for communicating with customers on:

- Product information
- Inquiries, contracts, or order handling (including amendments)
- Customer feedback (including customer complaints) [47]

One common result is the joint customer technical review. The IEEE 1028 Software Reviews provides further details on the review contents.

The last supplier responsibility from ISO 9001 Clause 7.1, Planning of Product Realization, is to plan and manage the processes needed for product realization. The most common product is the Software Project Management Plan (SPMP). IEEE Standard 1058, Software Project Management Plans, provides further details on specific SPMP contents.

Request For Proposal

The Customers Request For Proposal (RFP) is usually the first document to trigger the supply process. The RFP outline follows:

Project Title
Company Background
Project Description
Functional Requirements
Design Requirements
Technical and Infrastructure Requirements
Estimated Project Duration
Assumptions and Agreements
Submission Information
For Additional Information or Clarification
Basis for Award of Contract
Bidder's General Information
Anticipated Selection Schedule
Terms and Conditions

Request For Proposal (RFP) Guidance

The information presented below is based upon general recommendations provided in IEEE Standards 12207.0, 1220, and 12207.2.

Project Title. A carefully crafted "Request for Proposal" (RFP) is the key to getting the best quality services you require for your project. The RFP is your "official" statement to vendors about the products you require. Vendors typically try to respond, point by point, to your RFP when they make their proposals.

Company Background. Insert a concise paragraph outlining your company's background.

Project Description. Insert a summary of your project, including the problem/opportunity, goals/objectives, and any information that will help the vendors understand the need for the project. Include the context of this RFP. Be sure not to outline specific requirements in this section.

Functional Requirements. Insert an outline of all the functionality you would like your project to have with a short description. Differentiate between mandatory requirements and desirable requirements.

Design Requirements. Insert an outline of any requirements that pertain to the design of the project. Include any compatibility issues.

Technical and Infrastructure Requirements. Insert any technical or infrastructure-related requirements, such as a server or database configuration.

Estimated Project Duration. Insert the project start and estimated duration of the project or the required completion date. Also include a listing of expected of milestones and deliverables.

Assumptions and Agreements. Insert a list of any assumptions or agreements the product or the vendors must meet, such as ongoing communication channels or expected collaboration. Include the expected usage of the proposed product and prime contractor responsibility.

Submission Information. Insert the deadline for RFP submission, assumptions, and the address to submit the proposal to. Also state the confidentiality of evaluation and correction of errors.

For Additional Information or Clarification. Insert a list of contacts that are available to clarify any questions regarding the RFP. Decide whether to allow vendor suggestions or creativity and, thus, RFP resubmissions.

Basis for Award of Contract. Insert the outline of the list of the evaluation criteria that will be used to determine the best vendor proposal. It is not uncommon to list the weight that each criterion holds in relation to the others. Include here a description of required development process and revision cycles, standards of quality, quality measures, and so on.

Bidder's General Information. Insert the bidder's profile, customer base, financial details, technical capability (CMMI), quality certification (ISO 9001), conformance to industry standards, tender's qualifications, relevant services, any specific skills and experience, details of management and key personnel, and reference projects.

Anticipated Selection Schedule. Insert the schedule for your RFP selection process, including the closing date for receipt of proposal.

Terms and Conditions. Attach blank contract or spell out special conditions/requirements that vendors must meet. Also include buyer's responsibilities, such as payment schedule.

Joint Customer Technical Reviews

The information provided below is based upon the recommendations provided in IEEE Std 1028, IEEE Standard for Software Reviews. Technical reviews are an effective way to detect and identify problems early in the software development process. The criteria for technical reviews are presented here to help organizations define their review practices.

Introduction. Technical reviews are an effective way to evaluate a software product by a team of qualified personnel to identify discrepancies from specifications and standards. Although technical reviews identify anomalies, they may also provide the recommendation and examination of various alternatives. The examination need not address all aspects of the product and may only focus on a selected aspect of a software product.

Software requirements specifications, design descriptions, test and user documentation, installation and maintenance procedures, and build processes are all examples of items subject to technical reviews.

Responsibilities. The roles in support of a technical review are:

Decision Maker. The decision maker is the individual requesting the review and determines whether objectives have been met.

Review Leader. The review leader is responsible for the review and must perform all administrative tasks (to include summary reporting), ensure that the review is conducted in an orderly manner, and that the review meets its objectives.

Recorder. The recorder is responsible for the documentation of all anomalies, action items, decisions, and recommendations made by the review team.

Technical Staff. The technical staff should actively participate in the review and evaluation of the software product.

The following roles are optional and may also be established for the technical review:

Management Staff. Management staff may participate in the technical review for the purpose of identifying issues that will require management resolution.

Customer or User Representative. The role of the customer or user representative should be determined by the review leader prior to the review.

Input. Input to the technical review should include a statement of objectives, the software product being examined, existing anomalies and review reports, review procedures, and any standard against which the product is to be examined. Anomaly categories should be defined and available during the technical review. For additional information in support of the categorization of software product anomalies, refer to IEEE Std 1044 [13].

Authorization. All technical reviews should be defined in the SPMP. The plan should describe the review schedule and all allocation of resources. In addition to those technical reviews required by the SPMP, other technical reviews may be scheduled.

Preconditions. A technical review shall be conducted only when the objectives have been established and the required review inputs are available.

Procedures

Management Preparation. Managers are responsible for ensuring that all reviews are planned, that all team members are trained and knowledgeable, and that adequate resources are provided. They are also responsible for ensuring that all review procedures are followed.

Planning the Review. The review leader is responsible for the identification of the review team and their assignment of responsibility. The leader should schedule and announce the meeting, prepare participants for the review by providing them the required material, and collect all comments.

Overview of Review Procedures. The team should be presented with an overview of the review procedures. This overview may occur as a part of the review meeting or as a separate meeting.

Overview of the Software Product. The team should receive an overview of the software product. This overview may occur either as a part of the review meeting or as a separate meeting.

Preparation. All team members are responsible for reviewing the product prior to the review meeting. All anomalies identified during this prereview process should be presented to the review leader. Prior to the review meeting, the leader should classify all anomalies and forward these to the author of the software product for disposition.

The review leader is also responsible for the collection of all individual preparation times to determine the total preparation time associated with the review.

Examination. The review meeting should have a defined agenda that should be based upon the premeeting anomaly summary. Based upon the information presented, the team should determine whether the product is suitable for its intended use, whether it conforms to appropriate standards, and is ready for the next project activity. All anomalies should be identified and documented.

Rework/Follow-up. The review leader is responsible for verifying that the action items assigned in the meeting are closed.

Exit Criteria. A technical review shall be considered complete when all follow-up activities have been completed and the review report has been published.

Output. The output from the technical review should consist of the project being reviewed, a list of the review team members, a description of the review objectives, and a list of resolved and unresolved software product anomalies. The output should also include a list of management issues, all action items and their status, and any recommendations made by the review team.

Software Project Management Plan

IEEE Std 1058, IEEE Standard for Software Project Management Plans [15], and IEEE 12207.0, IEEE Standard for Life Cycle Processes [39], are effective instruments of the ISO 9001 requirements in support of software project planning. However, information regarding the measurement and measures required in support of software project planning activities needs to be added and stated explicitly. It is also important to note that though the requirements are addressed by the major headings of IEEE 1058, the details required to support ISO 9001 can be lost if each section is not carefully addressed while bearing the specific ISO 9001 project planning commitments requirements in mind. The modification of the recommended Software Project Management Plan (SPMP) table of contents to support the goals of ISO 9001 more directly is shown in Table 5-8.

Software Project Management Plan Document Guidance

The following provides section-by-section guidance in support of the creation of a SPMP. The SPMP should be considered to be a living document. The SPMP should change, in particular any associated schedules, and reflect any required change during the life cycle of a project. This guidance should be used to help define a management process and should reflect the actual processes and procedures of the implementing organization. Additional information is provided in the document template, *Software Project Management Plan.doc* that is located on the companion CD-ROM. Additional information is provided in Appendix C, Work Products, that describes the work breakdown structure, workflow diagram, and stakeholder involvement matrix work products.

> *Project Overview.* This paragraph should briefly state the purpose, scope, and objectives of the system and the software to which this document applies. It should describe the general nature of the system and software; summarize the history of system development, operation, and maintenance; identify the project sponsor, acquirer, user, developer, and support agencies; and identify current and planned operating sites. The project overview should also describe the relationship of this project to other projects, as appropriate, addressing any assumptions and constraints. This section should also provide a brief schedule and budget summary. This overview should not be construed as an official statement of product requirements. Reference to the official statement of product requirements should be provided in this subsection of the SPMP.
>
> *Project Deliverables.* This subsection of the SPMP should list all of the items to be delivered to the customer, the delivery dates, delivery locations, and quantities required to satisfy the terms of the project agreement. This list of project deliverables should not be construed as an official statement of project requirements.

Table 5-8. Software project management plan document outline

Title Page
Revision Page
Table of Contents
1. Introduction
    1.1   Project Overview
    1.2   Project Deliverables
    1.3   Document Overview
    1.4   Acronyms and Definitions
2. References
3. Project Organization
    3.1   Organizational Policies
    3.2   Process Model
    3.3   Organizational Structure
    3.4   Organizational Boundaries and Interfaces
    3.5   Project Responsibilities
4. Managerial Process
    4.1   Management Objectives and Priorities
    4.2   Assumptions, Dependencies, and Constraints
    4.3   Risk Management
    4.4   Monitoring and Controlling Mechanisms
    4.5   Staffing Plan
5. Technical Process
    5.1   Tools, Techniques, and Methods
    5.2   Software Documentation
    5.3   Project Support Functions
6. Work Packages
    6.1   Work Packages
    6.2   Dependencies
    6.3   Resource Requirements
    6.4   Budget and Resource Allocation
    6.5   Schedule
7. Additional Components

Document Overview. This subsection should summarize the purpose and contents of this document and describe any security or privacy considerations that should be associated with its use. This subsection of the SPMP should also specify the plans for producing both scheduled and unscheduled updates to the SPMP. Methods of disseminating the updates should be specified. This subsection should also specify the mechanisms used to place the initial version of the SPMP under change control and to control subsequent changes to the SPMP.

Acronyms and Definitions. This subsection should identify acronyms and definitions used within the project SPMP. The project SPMP should only list acronyms and definitions used within the SPMP.

References. This section should identify the specific references used within the project SPMP. The project SPMP should only contain references used within the SPMP.

Organizational Policies. This subsection of the SPMP should identify all organizational policies relative to the software project.

Process Model. This subsection of the SPMP should specify the (life cycle) software development process model for the project, describe the project organizational structure, identify organizational boundaries and interfaces, and define individual or stakeholder responsibilities for the various software development elements.

Organizational Structure. This subsection should describe the makeup of the team to be used for the project. All project roles, and stakeholders, should be identified as well as a description of the internal management structure of the project. Diagrams may be used to depict the lines of authority, responsibility, and communication within the project.

Organizational Boundaries and Interfaces. This subsection should describe the limits of the project including, any interfaces with other projects or programs, the application of the program's SCM and SQA (including any divergence from those plans), and the interface with the project's customer. This section should describe the administrative and managerial boundaries between the project and each of the following entities: the parent organization, the customer organization, subcontracted organizations, or any other organizational entities that interact with the project. In addition, the administrative and managerial interfaces of the project-support functions, such as configuration management, quality assurance, and verification should be specified in this subsection.

Project Responsibilities. This subsection should describe the project's approach through a description of the tasks required to complete the project (e.g., requirements → design → implementation → test) and any efforts (update documentation, etc.) required to successfully complete the project. It should state the nature of each major project function and activity, and identify the individuals, or stakeholders, who are responsible for those functions and activities.

Managerial Process. This section should specify management objectives and priorities; project assumptions, dependencies, and constraints; risk management techniques; monitoring and controlling mechanisms to be used; and the staffing plan.

Management Objectives and Priorities. This subsection should describe the philosophy, goals, and priorities for management activities during the project. Topics to be specified may include, but are not limited to, the frequency and mechanisms of reporting to be used; the relative priorities among requirements, schedule, and budget for this project; risk management procedures to be followed; and a statement of intent to acquire, modify, or use existing software.

Assumptions, Dependencies, and Constraints. This subsection should state the assumptions on which the project is based, the external events the project is dependent upon, and the constraints under which the project is to be conducted.

Risk Management. This subsection should identify the risks for the project. Completed risk management forms should be maintained and tracked by the project leader with associated project information. These forms should be reviewed at weekly staff meetings. Risk factors that should be considered include contractual risks, technological risks, risks due to size and complexity of the project, risks in personnel acquisition and retention, and risks in achieving customer acceptance of the product.

Monitoring and Controlling Mechanisms. This subsection should define the reporting mechanisms, report formats, information flows, review and audit mechanisms, and other tools and techniques to be used in monitoring and controlling adherence to the SPMP. A typical set of software reviews is listed in Appendix C. Project monitoring should occur at the level of work packages. The relationship of monitoring and controlling mechanisms to the project-support functions should be delineated in this subsection. This subsection should also describe the approach to be followed for providing the acquirer or its authorized representative access to developer and subcontractor facilities for review of software products and activities.

Staffing Plan. This subsection should specify the numbers and types of personnel required to conduct the project. Required skill levels, start times, duration of need, and methods for obtaining, training, retaining, and phasing out of personnel should be specified.

Technical Process. This section should specify the technical methods, tools, and techniques to be used on the project. In addition, the plan for software documentation should be specified, and plans for project support functions such as quality assurance, configuration management, and verification and validation may be specified.

Tools, Techniques, and Methods. This subsection of the SPMP should specify the computing system(s), development methodology(s), team structures(s), programming language(s), and other notations, tools, techniques, and methods to be used to specify, design, build, test, integrate, document, deliver, modify or maintain or both (as appropriate) the project deliverables.

   This subsection should also describe any tools (compilers, CASE tools, and project management tools), any techniques (review, walk-through, inspection, prototyping) and the methods (object-oriented design, rapid prototyping) to be used during the project.

Software Documentation. This subsection should contain, either directly or by reference, the documentation plan for the software project. The documentation plan should specify the documentation requirements, and the milestones, baselines, reviews, and sign-offs for software documentation. The documentation plan may also contain a style guide, naming conventions, and documentation formats. The documentation plan should provide a summary of the schedule and resource requirements for the documentation effort. IEEE Std for Software Test Documentation (IEEE Std 829-1998) [5] provides a standard for software test documentation.

Project Support Functions. This subsection should contain, either directly or by reference, plans for the supporting functions for the software project. These functions may include, but are not limited to, configuration management, software quality assurance, and verification and validation.

Work Packages. This section of the SPMP should specify the work packages, identify the dependency relationships among them, state the resource requirements, provide the allocation of budget and resources to work packages, and establish a project schedule.

The work packages for the activities and tasks that must be completed in order to satisfy the project agreement must be described in this section. Each work package should be uniquely identified; identification may be based on a numbering scheme and descriptive titles. A diagram depicting the breakdown of activities into subactivities and tasks may be sued to depict hierarchical relationships among work packages.

Dependencies. This subsection should specify the ordering relations among work packages to account for interdependencies among them and dependencies on external events. Techniques such as dependency lists, activity networks, and the critical path may be used to depict dependencies.

Resource Requirements. This subsection should provide, as a function of time, estimates of the total resources required to complete the project. Numbers and types of personnel, computer time, support software, computer hardware, office and laboratory facilities, travel, and maintenance requirements for the project resources are typical resources that should be specified.

Budget and Resource Allocation. This subsection should specify the allocation of budget and resources to the various project functions, activities, and tasks. Defined resources should be tracked.

Schedule. This subsection should be used to capture the project's schedule, including all milestones and critical paths. Options include Gantt charts (Milestones Etc.™ or Microsoft Project™), Pert charts, or simple time lines.

Additional Components. This section should address additional items of importance on any particular project. This may include subcontractor management plans, security plans, independent verification and validation plans, training plans, hardware procurement plans, facilities plans, installation plans, data conversion plans, system transition plans, or product maintenance plans.

DEVELOPMENT

The development process contains the activities and tasks of the developer. The developer manages the development process at the project level following the management process, infrastructure process, and tailoring process. Also, the developer manages the process at the organizational level following the improvement process and the training process. Finally the developer performs the supply process if it is the supplier of developed software products. Table 5-9 provides a list of the Development process objectives:

The development process is the largest of the 17 processes in IEEE 12207. The development activities are:

- Process implementation
- System requirements analysis
- System architectural design
- Software requirements analysis
- Software architectural design
- Software detailed design
- Software coding and testing
- Software integration
- Software qualification testing
- System integration
- System qualification testing
- Software installation
- Software acceptance support

Depending upon the type of contract, the development process begins with process implementation and continues through to the customer acceptance.

ISO 9001 Goals

ISO 9001 identifies the requirements associated with development as "product realization." All products must be planned. That is, the development of processes associated with product realization should be documented and these processes should reflect the organization's actual method of operation. ISO 9001 describes the requirements in support of the IEEE 12207 Development Process in Clause 7.3, Design and Development. These requirements cover the processes, supporting information, and verification activities required in support of the development process.

As ISO 9001 is generic for all product and services and written at a higher level than the more detailed IEEE 12207, ISO 9001 has fewer subclauses than IEEE 12207:

7.3.1 Design and Development Planning

7.3.2 Design and Development Inputs

7.3.3 Design and Development Outputs

7.3.4 Design and Development Review

Table 5-9. Development process objectives

a) Develop requirements of the system that match the customer's stated and implied needs.

b) Propose an effective solution that identifies the main elements of the system.

c) Allocate the defined requirements to each of those main elements.

d) Develop a system release strategy.

e) Communicate the requirements, proposed solution, and their relationships to all affected parties.

f) Define the requirements allocated to software components of the system and their interfaces to match the customer's stated and implied needs.

g) Develop software requirements that are analyzed, correct, and testable.

h) Understand the impact of software requirements on the operating environment.

i) Develop a software release strategy.

j) Approve and update the software requirements, as needed.

k) Communicate the software requirements to all affected parties.

l) Develop an architectural design.

m) Define internal and external interfaces of each software component.

n) Establish traceability between system requirements and design and software requirements, between software requirements and software design, and between software requirements and tests.

o) Define verification criteria for all software units against the software requirements.

p) Produce software units defined by the design.

q) Accomplish verification of the software units against the design.

r) Develop an integration strategy for software units consistent with the release strategy.

s) Develop acceptance criteria for software unit aggregates that verify compliance with the software requirements allocated to the units.

t) Verify software aggregates using the defined acceptance criteria.

u) Verify integrated software using the defined acceptance criteria.

v) Record the results of the software tests.

w) Develop a regression strategy for retesting aggregates, or the integrated software, should a change in components be made.

x) Develop an integration plan to build system unit aggregates according to the release strategy.

y) Define acceptance criteria for each aggregate to verify compliance with the system requirements allocated to the units.

z) Verify system aggregates using the defined acceptance criteria.

aa) Construct an integrated system demonstrating compliance with the system requirements (functional, nonfunctional, operations and maintenance).

ab) Record the results of the system tests.

ac) Develop a regression strategy for retesting aggregates or the integrated system should a change in components be made.

ad) Identify transition concerns, such as availability or work products, availability of system resources to resolve problems and adequately test before fielding corrections, maintainability, and assessment of transitioned work products.

7.3.5 Design and Development Verification

7.3.6 Design and Development Validation

7.3.7 Control of Design and Development Changes

The development process, in conformance to ISO 9001, can leverage several IEEE standards:

- IEEE Std 829, Standard for Software Test Documentation
- IEEE Std. 830, Recommended Practice for Software Requirements Specifications
- IEEE Std 1008, Standard for Software Unit Testing
- IEEE Std 1012, Standard for Software Verification and Validation Plans
- IEEE Std. 1016, Recommended Practice for Software Design Descriptions
- IEEE Std 1063, Standard for Software User Documentation
- IEEE Std. 1074 Standard for Developing a Software Project Life Cycle Process
- IEEE Std 1220, Standard for Application and Management of the Systems Engineering Process
- IEEE Std 1233, Guide to Developing System Requirements Specifications
- IEEE Std. 1320.1, Standard for Functional Modeling Language—Syntax and Semantics for IDEF0
- IEEE Stds1420.1, 1420.1a, and 1420.1b Software Reuse—Data Model for Reuse Library Interoperability
- IEEE Std. 1471, Recommended Practice for Architectural Description of Software Intensive Systems
- IEEE/EIA Std. 12207.0, Standard for Information Technology—Software Life Cycle Processes
- IEEE/EIA Std. 12207.1, Standard for Information Technology—Software Life Cycle Processes—Life cycle data,

The next sections will address each development activity and the related project documentation supported by the IEEE standards. Unless stipulated in the contract, the developer should define the software life cycle model for the project. IEEE Std. 1074, Standard for Developing a Software Project Life Cycle Process, is specifically constructed for this development activity of process implementation.

ISO 9001 describes requirements in support of the IEEE 12207 development process for process implementation activity in Clause 7.1, Planning of Product Realization. This requires planning the necessary processes, documents, and resources, followed up by the keeping of records as evidence that the processes and resulting product meet requirements. The process implementation should result in a definition of what products are to be produced, who is to produce them, and when they are to be produced and verified.

Clause 7.3.1, Design and Development Planning, also applies to process implementation, as the software project life cycle implementation must determine the stages of design and development and their appropriate review, verification, and validation activities for each stage. Development responsibilities and authorities are included.

ISO 9001 describes requirements in support of the IEEE 12207 development process for the system requirements analysis activity in Clause 7.3.2, Design and Development Inputs. This requires that each software requirement must be determined and that records must be maintained. Each requirement must be reviewed for sufficiency and completeness. Any incomplete, ambiguous, or conflicting requirement must be resolved. The requirements should include:

- Functional and performance requirements
- Applicable statutory and regulatory requirements
- Applicable information derived from similar designs
- Requirements essential for design and development [47]

ISO 9001 describes requirements in support of the IEEE 12207 development process for system architectural design activity in Clause 7.3.2, Design and Development Inputs. This requires software requirement inputs to be determined and for records to be maintained. These inputs must be reviewed for adequacy. Any incomplete, ambiguous, or conflicting requirements must be resolved.

ISO 9001 describes requirements in support of the IEEE 12207 development process for Software requirements analysis activity in Clause 7.3.2, Design and Development Inputs. This requires software requirement inputs to be determined and records maintained. These inputs must be reviewed for adequacy. Any incomplete, ambiguous, or conflicting requirements must be resolved.

ISO 9001 also describes the requirements in support of IEEE 12207 development processes for the software architectural design activities in Clause 7.3.3, Design and Development Outputs. This requires that all design and development outputs must be documented to enable verification against the inputs to the design and development process. The design and development outputs must:

- Meet design and development input requirements
- Provide information for purchasing, production, and service
- Contain or reference product acceptance criteria
- Define essential characteristics for safe and proper use [47]

ISO 9001 describes requirements in support of the IEEE 12207 development process for software detailed design activity in Clause 7.3.3, Design and Development Outputs. This requires design and development outputs to be documented to enable verification against the inputs to the design and development process.

ISO 9001 describes requirements in support of the IEEE 12207 development process for software coding and testing activity in Clause 7.3.3, Design and Development Outputs. This requires design and development outputs to be documented to enable verification against the inputs to the design and development process.

ISO 9001 describes requirements in support of the IEEE 12207 development process for software integration activity in Clause 7.3.5, Design and Development Verification. This requires test planning and execution to be performed in accordance with planned SLC to ensure that the output meets the design and development input requirements. The results of the verification and subsequent follow-up actions must be maintained.

ISO 9001 describes requirements in support of the IEEE 12207 development process for software qualification testing activity in Clause 7.3.6, Design and Development Validation. This requires reviews, test planning, and execution to be performed in accordance with the planned SLC to confirm the resulting software product is capable of meeting the requirements for its specified application or intended use. When practical, the validation must be completed before software delivery or implementation. The results of the verification and subsequent follow-up actions must be maintained.

ISO 9001 describes requirements in support of the IEEE 12207 development process for system integration activity in Clause 7.3.5, Design and Development Verification. This requires test planning and execution to be performed in accordance with planned SLC to ensure that the output meets the design and development input requirements. The results of the verification and subsequent follow-up actions must be maintained.

ISO 9001 describes requirements in support of the IEEE 12207 development process for system qualification testing activity in Clause 7.3.6, Design and Development Validation. This requires reviews, test planning and execution to be performed in accordance with the planned SLC to confirm that the resulting software product is capable of meeting the requirements for its specified application or intended use. When practical, the validation must be completed before software delivery or implementation. The results of the verification and subsequent follow-up actions must be maintained.

ISO 9001 describes requirements in support of the IEEE 12207 development process for software installation activity in Clause 7.3.6, Design and Development Validation, and in Clause 7.5.1, Control of Production and Service Provision. Clause 7.3.6 requires reviews, test planning, and execution to be performed in accordance with the planned SLC to confirm that the resulting software product is capable of meeting the requirements for its specified application or intended use. The results of the verification and subsequent follow-up actions must be maintained. Clause 7.5.1 requires planning and execution of software production and service provision under controlled conditions, to include:

- Availability of software product characteristics information
- Availability of work instructions
- Use of suitable equipment
- Availability and use of monitoring and measuring devices

- Implementation of monitoring and measurement activities
- Implementation of release, delivery, and postdelivery activities

Related to this is Clause 7.5.5, Preservation of Product, which requires that the software product be preserved to the customer requirements during internal processing and delivery to the intended destination. This preservation shall include identification, handling, packaging, storage, and protection.

ISO 9001 describes requirements in support of the IEEE 12207 development process for software acceptance support activity in Clause 7.3.6, Design and Development Validation, and in 7.5.1, Control of Production and Service Provision.

These clauses, when combined, distill the development process. Much is required in order to successfully support these ISO clauses. The information provided in this chapter describes much of the documentation that may be used to support the IEEE 12207 process development process in ISO 9001, Clause 7.1, Planning of Product Realization, Clause 7.3, Design and Development, and Clause 7.5.1, Control of Production and Service Provision.

System Requirements Analysis

When the software product is part of a system, then the SLC will specify the system requirements analysis activity. This analysis will result in the system requirements specification, which describes functions and capabilities of the system; business, organizational and user requirements; safety, security, human-factors engineering (ergonomics), interface requirements, operational, and maintenance requirements; and design constraints and qualification requirements.

System Requirements Specification

The information provided here in support of requirements management is designed to facilitate the definition of a system requirements specification. This information was developed using IEEE Std 1233, IEEE Guide for Developing System Requirements Specifications (SysRS) [25], which has been adapted to support ISO 9001 requirements. The modification of the recommended system requirements specification (SysRS) table of contents to support the goals of ISO 9001 methodology more directly could look like the one Table 5-10.

System Requirements Specification Document Guidance

The following provides section-by-section guidance in support of the creation of a SysRS. This guidance should be used to help establish a requirements baseline and should reflect the actual processes and procedures of the implementing organization. Additional information is provided in the document template, *System Requirements Specification.doc,* which is located on the companion CD-ROM.

Table 5-10. System requirements specification document outline

Title Page
Revision Page
Table of Contents
1. Introduction
    1.1   System Purpose
    1.2   System Scope
    1.3   Definitions, Acronyms, and Abbreviations
    1.4   References
    1.5   System Overview
2. General System Description
    2.1   System Context
    2.2   System Modes and States
        2.2.1   Configurations
    2.3   Major System Capabilities
    2.4   Major System Conditions
    2.5   Major System Constraints
    2.6   User Characteristics
    2.7   Assumptions and Dependencies
    2.8   Operational Scenarios
3. System Capabilities, Conditions, and Constraints
    3.1   Physical
        3.1.1   Construction
        3.1.2   Durability
        3.1.3   Adaptability
        3.1.4   Environmental Conditions
    3.2   System Performance Characteristics
        3.2.1   Load
        3.2.2   Stress
        3.2.3   Contention
        3.2.4   Availability
    3.3   System Security and Safety
    3.4   Information Technology Management
    3.5   System Operations
        3.5.1   System Human Factors
        3.5.2   System Usability
        3.5.3   Internationalization
        3.5.4   System Maintainability
        3.5.5   System Reliability
    3.6   Policy and Regulation
    3.7   System Life Cycle Sustainment
4. System Interfaces
5. Specific Requirements
6. Traceability Matrix

Introduction. Explain the purpose and scope of the project system requirements specification (SysRS), as well as, provide clarification of definitions, acronyms, and references. This section should also provide an overview of the project.

System Purpose. Explain the purpose for writing the SysRS for this project and describe the intended audience for the SysRS. (Note this maybe aligned with the ConOps document, System Overview.)

System Scope. Identify the system products to be produced, by name; explain what the system products will, and will not, do, and describe the application of the system being specified, including all relevant goals, objectives, and benefits from producing the system. (Note this maybe aligned with ConOps document, System Overview.)

Definitions, Acronyms, and Abbreviations. Provide the definitions of all terms, acronyms, and abbreviations required to properly interpret the SysRS. (Note this maybe aligned with ConOps document, Definitions and Acronyms.)

Key References. List all references used within the SysRS.

System Overview. Describe what the rest of the SysRS contains as it relates to the systems and software components effort. (Note this maybe aligned with ConOps document, System Overview.) It should also explain how the document is organized.

General System Description. Describe the general factors that affect the product and its requirements. (Note this maybe aligned with ConOps document, Proposed System.)

System Context. Include diagrams and narrative to provide an overview of the context of the system, defining all significant interfaces crossing the system's boundaries.

System Modes and States. Include diagrams and narrative to provide usage/operation modes and transition states (Note this maybe aligned with ConOps document, Modes of Operation.)

Configurations. Describe typical system configurations that meet various customers' needs.

Major System Capabilities. Provide diagrams and accompanying narrative to show major capability groupings of the requirements.

Major System Conditions. Show major conditions, relative to the attributes of the major capability groupings.

Major System Constraints. Show major constraints, relative to the boundaries of the major capability groupings.

User Characteristics. Identify each type of user of the system (by function, location, and type of device), the number in each group, and the nature of their use of the system. (Note this maybe aligned with ConOps document, User Classes.)

Assumptions and Dependencies. Address all assumptions and dependencies that impact the system resulting from the SysRS. It is important to tie assumptions to assessments of impact. This subsection should be the source for recognizing the impact of any changes to the assumptions or dependencies on the SysRS and the resulting system. This section can highlight unresolved requirement issues and should be recorded on the Project Manager's Open Issues List.

Operational Scenarios. Provide descriptive examples of how the system will be used. Scenarios may also be used to describe what the system will not do. Scenarios should be used to help readers understand how system functionality will support operational requirements. (Note this maybe aligned with ConOps document, Operational Scenarios.)

System Capabilities, Conditions, and Constraints. Describe the next levels of system structure. System behavior, exception handling, manufacturability, and deployment should be covered under each capability, condition, and constraint.

Physical

Physical Construction. Include the environmental (mechanical, electrical, chemical) characteristics of where the system will be installed.

Physical Dependability. Include the degree to which a system component is operable and capable of performing its required function at any (random) time, given its suitability for the mission and whether the system will be available and operable as many times and as long as needed. Examples of dependability measures are availability, interoperability, compatibility, reliability, repeatability, usage rates, vulnerability, survivability, penetrability, durability, mobility, flexibility, and reparability.

Physical Adaptability. Include how to address growth, expansion, capability, and contraction.

Environmental Conditions. Include environmental conditions to be encountered by the system. The following subjects should be considered for coverage: natural environment (wind, rain, and temperature); induced environment (motion, shock, and noise); and electromagnetic signal environment.

System Performance Characteristics. Include the critical performance conditions and their associated capabilities. Consider performance requirements for the operational phases and modes. Also consider endurance capabilities, minimum total life expectancy, operational session duration, planned utilization rate, and dynamic actions or changes.

Load. Describe the range of work or transaction loads the system will operate at or process in one time period.

Stress. Describe the extreme but valid conditions for system work or transaction loads.

Contention. Identify any sharing of computer resources whereby two or more processes can access the computer's processor simultaneously.

Availability. Describe the range of operational availability or the actual availability of a system under logistics constraints.

System Security and Safety. Cover both the facility that houses the system and operational security requirements, including requirements of privacy, protection factors, and safety.

Information Technology Management. Describe the various stages of information processing from production to storage and retrieval to dissemination toward the better working of the user organization.

System Operations. Describe how the system supports the user needs and the needs of the user community.

    System Human Factors. Include references to applicable documents and specify any special or unique requirements for personnel and communications and personnel/equipment interactions.

    System Usability. Describe the special or unique usability requirements for personnel and communications and personnel/equipment interactions.

    Internationalization. For each major global customer segment, describe the special or unique internationalization requirements for personnel and communications and personnel/equipment interactions.

    System Maintainability. Describe, in quantitative terms, the requirements for the planned maintenance, support environment, and continued enhancements.

    System Reliability. Specify, in quantitative terms, the reliability conditions to be met. Consider including the reliability apportionment model to support allocation of reliability values assigned to system functions for their share in achieving desired system reliability.

    Policy and Regulation. Describe relevant organizational policies that will affect the operation or performance of the system. Include any external regulatory requirements or constraints imposed by normal business practices.

    System Life Cycle Sustainment. Describe quality activities and measurement collection and analysis performed to help sustain the system through its expected life cycle.

System Interfaces. Include diagrams and narrative to describe the interfaces among different system components and their external capabilities, including all users, both human and other systems. Include interface interdependencies or constraints.

Specific Requirements. This section should contain all details that the system/software developer needs to create a design. The details within this section should be defined as individual, specific requirements with sufficient cross-referencing back to related discussions in the Introduction and General Description sections above. Each specific requirement should be stated such that its achievement can be objectively verified by a prescribed method. Each requirement should enumerated for tracking. The method of verification will be shown in a requirements traceability matrix in Appendix A of the SySRS.

Traceability Matrix. This is a comprehensive baseline of all system requirements described within the SySRS. This list should be used as the basis for all requirements management, system and allocated software requirements, design, development, and test.

Software Requirements Analysis

The resulting software requirements specifications should include the quality characteristics, functional and capability characteristics, external interfaces, qualification, safety, security, human-factors engineering, data definitions, installation and acceptance, user documentation, user operations, and user maintenance.

Software Requirements Specification

The information provided here in support of requirements management is designed to facilitate the definition of a software requirements specification. This information was developed using IEEE Std 830, IEEE Recommended Practice for Software Requirements Specifications [6] which has been adapted to support ISO 9001 requirements. The modification of the recommended Software Requirements Specification table of contents to support the goals of ISO 9001 more directly is shown in Table 5-11.

Software Requirements Specification Document Guidance

The following provides section-by-section guidance in support of the creation of a software requirements specification. This guidance should be used to help establish a requirements baseline and should reflect the actual processes and procedures of the implementing organization. Additional information is provided in the document template, *Software Requirements Specification.doc,* which is located on the companion CD-ROM.

Title Page. The title page should include the names and titles of all document approval authorities.

Revision Sheet. The revision sheet should provide the reader with a list of cumulative document revisions. The minimum information recorded should include the revision number, the document version number, the revision date, and a brief summary of changes.

Introduction. This section should explain the purpose and scope of the project software requirements specification (SRS), as well as provide clarification of

Table 5-11. Software requirements
specification document outline

Title Page
Revision Page
Table of Contents
1. Introduction
   1.1 Software Purpose
   1.2 Software Scope
   1.3 Definitions, Acronyms, and Abbreviations
   1.4 References
   1.5 Software Overview
2. Overall Description
   2.1 Product Perspective
   2.2 Product Functions
   2.3 Environmental Conditions
   2.4 User Characteristics
   2.5 External Interfaces
   2.6 Constraints
      2.6.1 Safety
      2.6.2 Security
      2.6.3 Human factors
   2.7 Assumptions and Dependencies
3. Requirements Management
   3.1 Resources and Funding
   3.2 Reporting Procedures
   3.3 Training
4. Specific Requirements
5. Acceptance
6. Documentation
7. Maintenance
Appendixes
Traceability Matrix
Index

definitions, acronyms, and references. This section should also provide an overview of the project.

Software Purpose. This subsection should explain the purpose for writing an SRS for this project and describe the intended audience for the SRS.

Software Scope. This subsection should identify the software products to be produced, by name; explain what the software products will, and if necessary, will not do; and describe the application of the software being specified, including all relevant goals, objectives, and benefits from producing the software.

Definitions, Acronyms, and Abbreviations. This section should provide the definitions of all terms, acronyms, and abbreviations required to properly interpret the SRS.

Key References. This subsection should list all references used within the SRS.

Software Overview. This subsection should describe what the rest of the SRS contains as it relates to the software effort. It should also explain how the document is organized.

Overall Description. This section should describe the general factors that affect the product and its requirements.

Product Perspective. This subsection should put the product into perspective with other related products or projects. If the product to be produced from this SRS is totally independent, it should be clearly stated here. If the product to be produced from this SRS is part of a larger system, then this subsection should describe the functions of each component of the larger system or project and identify the interfaces between this product and the remainder of the system or project. This subsection should identify all principal external interfaces for this software product (Note: descriptions of the interfaces will be contained in another part of the SRS). The product perspective requirements may be documented under Product Packaging Information, shown in Appendix C.

Product Functions. This subsection should provide a summary of the functions to be performed by the software produced as a result of this SRS. Functions listed in this section should be organized in a way that will make it understandable to the intended audience of the SRS. (Note: this subsection is an overview; details of the specific requirements will be contained in Section 4 of the SRS.)

Environmental Conditions. This subsection should provide a summary of the environment in which the software must operate. (Note: this subsection is an overview; details of the specific requirements will be contained in Section 4 of the SRS.)

User Characteristic. This subsection should describe the general characteristics of the eventual users of the product that will affect the specific requirements. Eventual users of the product will include end-product customers, operators, maintainers, and systems people as appropriate. For any users that impact the requirements, characteristics such as education, skill level, and experience levels will be documented within this subsection as they impose constraints on the product.

External Interfaces. This subsection should describe all required external interface requirements. This should include references to any existing interface control documentation. (Note: this subsection is an overview; details of the specific requirements will be contained in Section 4 of the SRS.)

Constraints. This subsection should provide a list of the general constraints imposed on the system that may limit designer's choices. Constraints may come from regulations or policy, hardware limitations, interfaces to other applications, parallel operations, audit functions, control functions, higher-order language requirements, communication protocols, criticality of the applications or safety, and security restrictions. This subsection should not describe the implementation of the constraints, but should provide a source for understanding why certain constraints will be expected from the design.

Assumptions and Dependencies. This subsection should list all assumptions and dependencies that impact the product of the software resulting from the SRS. This subsection should be the source for recognizing the impact of any changes to the assumptions or dependencies on the SRS and resulting software. This subsection can highlight unresolved requirement issues and should be recorded on the Project Manager's Open Issues List.

Requirements Management. This section should provide an overview of the requirements management process and any associated procedures. References to existing organizational requirements management plans may be used.

Resources and Funding. This subsection should identify individuals responsible for requirements management. If this information were provided in an existing requirements management plan, a reference to this plan would be provided here.

Reporting Procedures. This subsection should describe all reporting procedures associated with requirements management activities. Details should be provided for the reporting procedures and reporting schedule. If this information is provided in an organizational software requirements management plan, or if this information is provided in the software project management plan, then a reference to these plans should be provided.

Training. This subsection should either reference an organizational training plan, which describes the method and frequency of requirements elicitation, documentation, and management training, or provide the details for all associated training.

Specific Requirements. This section should contain all details that the software developer needs to create a design. The details within this section should be defined as individual, specific requirements with sufficient cross-referencing back to related discussions in the Introduction and General Description sections above. Each specific requirement should be stated such that its achievement can be objectively verified by a prescribed method. The method of verification will be shown in a requirements traceability matrix in Appendix A of the SRS.

Acceptance. This section should describe the acceptance process, referencing any associated documentation (e.g., software test plan).

Documentation. This section should list any additional supporting process or project documentation not previously referenced.

Maintenance. This section is required if the system must meet requirements for continued operation and enhancement.

Traceability Matrix. This is a comprehensive baseline of all software requirements described within the SRS. This list should be used as the basis for all requirements management, software design, development, and test. Please refer to Appendix C, Work Products, for additional information in support of a requirements traceability matrix.

Software Design Document

The following information is primarily based on IEEE Std 1016, IEEE Recommended Practice for Software Design Descriptions [11], and IEEE 12207.1[39], Standard for Information Technology—Software Life Cycle Processes—Life Cycle Data. IEEE Std 1471, IEEE Recommended Practice for Architectural Description of Software Intensive Systems [33]; IEEE Std 1320.1, IEEE Standard for Functional Modeling Language—Syntax and Semantics for IDEF0 [26]; and IEEE Stds1420.1, 1420.1a, and 1420.1b, Software Reuse—Data Model for Reuse Library Interoperability [29], were also used as source documents for the development of this information. These standards have been adapted to support ISO 9001 requirements. Table 5-12 provides a sample document outline supporting ISO 9001 requirements. The resulting Software Design Document (SDD) describes the detailed design for each software component of the software item. The software components must be refined into lower levels containing software units that can be coded, compiled, and tested. It must be ensured that all the software requirements are allocated from the software components to software units.

Software Design Document Guidance

The following provides section-by-section guidance in support of the creation of an SDD. This guidance should be used to help define associated design decisions and should reflect the actual requirements of the implementing organization. Additional information is provided in the document template, *Software Design Document.doc,* which is located on the companion CD-ROM.

> *Introduction.* As stated in IEEE 1016, "The SDD shows how the software system will be structured to satisfy the requirements identified in the software requirements specification. It is a translation of requirements into a description of the software structure, software components, interfaces, and data necessary for implementation. In essence, the SDD becomes a detailed blueprint for the implementation activity."
>
> This section should explain the purpose and scope of project software design, as well as, provide clarification of definitions, acronyms, and references. This section should also provide an overview of the project.
>
> *Purpose.* This subsection should explain the purpose for writing an SDD for this project and describe the intended audience for the SDD.

Table 5-12. Software design document outline

Title Page
Revision Page
Table of Contents
1. Introduction
   1.1 Purpose
   1.2 Scope
   1.3 Definitions, Acronyms, and Abbreviations
   1.4 References
2. Design Overview
   2.1 Background Information
   2.2 Alternatives
3. User Characteristics
4. Requirements and Constraints
   4.1 Performance Requirements
   4.2 Security Requirements
   4.3 Design Constraints
5. System Architecture
6. Detailed Design
   6.1 Description for Component N
   6.2 Component N Interface Description
7. Data Architecture
   7.1 Data Analysis
   7.2 Output Specifications
   7.3 Logical Database Model
   7.4 Data Conversion
8. Interface Requirements
   8.1 Required Interfaces
   8.2 External System Dependencies
9. User Interface
   9.1 Module [X] Interface Design
   9.2 Functionality
10. Nonfunctional Requirements
11. Traceability Matrix

Scope. This subsection should:
- Identify the software products to be produced, by name.
- Explain what the software products will and, if necessary, will not, do.
- Describe the application of the software being specified, including all relevant goals, objectives, and benefits from producing the software.

Provide a description of the dominant design methodology. Provide a brief overview of the product architecture. Briefly describe the external systems with which this system must interface. Also explain how this document might evolve throughout the project lifecycle.

Definitions, Acronyms, and Abbreviations. This subsection should provide the definitions of all terms, acronyms, and abbreviations required to properly interpret the SDD.

References. This subsection should list all references used within the SDD. All relationships to other plans and policies should be described as well as existing design standards.

Design Overview

Background Information. This subsection should briefly present background information relevant to the development of the system design. All stakeholders should be identified and their contact information provided. A description of associated project risks and issues may also be presented along with assumptions and dependencies critical to project success. Describe the business processes that will be modeled by the system.

Alternatives. All design alternatives considered, and the rationale for nonacceptance, should be briefly addressed in this subsection. See Appendix C for the Make/Buy/Mine/Commission Decision Matrix and Alternative Solution Screening Criteria Matrix.

User Characteristics. Identify the potential system users. Specify the levels of expertise needed by the various users and indicate how each user will interact with the system. Describe how the system design will meet specific user requirements.

Requirements and Constraints

Performance Requirements. This subsection should describe how the proposed design will ensure that all associated performance requirements will be met.

Security Requirements. This subsection should describe how the proposed design will ensure that all associated security requirements will be met. List any access restrictions for the various types of system users. Describe any access code systems used in the software. Identify any safeguards that protect the system and its data. Specify communications security requirements.

Design Constraints. This subsection should describe how requirements that place constraints on the system design would be addressed. List all dependencies and limitations that may affect the software. Examples include budget and schedule constraints, staffing issues, and availability of components.

System Architecture. This section should provide a description of the architectural design. See Appendix C for Architecture Design Success Factors and Pitfalls. All entities should be described as well as their interdependent relationships. A top-level diagram may be provided. See Appendix C for Unified Modeling Language (UML), an example showing inheritance, aggregation, and reference relationships.

Detailed Design. Decompose the system into design components that will interact with and transform data to perform the system objectives. Assign a unique name to each component, and group these components by type, for example, class, object, and procedure. Describe how each component satisfies system requirements. In user terminology, specify the inputs, outputs, and transformation rules for each component. Depict how the components depend on each other. Each component should be described as shown by the following list of suggested headers:

- Description for Component n
- Processing Narrative for Component n
- Component n Interface Description
- Component n Processing Detail

Data Architecture. This section should describe the data structures to be used in support of the implementation. If these include databases, define the table structure, including full field descriptions, relationships, and critical database objects. Graphical languages are appropriate. This information is often provided in a separate Database Design Document. If this is the case, simply refer to this document and omit the remainder of this section.

Data Analysis. A brief description of the procedures used in support of data analysis activities should be described in this subsection. Any analysis of the data that resulted in a change to the system design, or that impacted the system design, should be noted.

Output Specifications. All designs supporting requirements for system outputs should be described in this subsection. These may include designs to support reporting, printing, e-mail, and so on.

Logical Database Model. Identify specific data elements and logical data groupings that are stored and processed by the design components in the Detailed Design. Outline data dependencies, relationships, and integrity rules in a data dictionary. Specify the format and attributes of all data elements or data groupings. A logical model of data flow, depicting how design elements transform input data into outputs, should be developed and presented here.

Data Conversion. This subsection should describe all design requirements in support of data conversion activities. This may be covered in a separate data migration plan, but if it is not, it should be documented in the design documentation. The migration and validation of any converted legacy data should be described.

Interface Requirements

Required Interfaces. This subsection should describe all interfaces required in support of hardware and software communications. If an Interface Control Document was developed along with the Software Requirements Specification it should be referenced here. The effectiveness of how the system design addresses relevant interface issues should be discussed. Specify how the product will interface with other systems. For each interface, describe the inputs and outputs for the interacting systems. Explain how data is formatted for transmission and validated upon arrival. Note the frequency of data exchange.

External System Dependencies. This subsection should provide a description of all external system dependencies. A diagram may be used to provide a description of the system with each processor and device indicated.

User Interface. Describe the user interface and the operating environment, including the menu hierarchy, data entry screens, display screens, online help, and system messages. Specify where in this environment the necessary inputs are made, and list the methods of data outputs, for example, printer, screen, or file.

Module [X] Interface Design. This subsection should contain screen images and a description of all associated design rules.

Functionality. All objects and actions should be described in this subsection. Any reuse of existing components should be identified.

Nonfunctional Requirements. This section should address design items associated with nonfunctional requirements relating to system performance, security, licensing, language, or other related items.

Traceability Matrix. It is necessary to show traceability throughout the product life cycle. All design items should be traceable to original system requirements. This can be annotated in the Requirements Traceability Matrix, the Design Review Document, or in the Configuration Management System.

Interface Control Document

The following information is based on IEEE Std 830, IEEE Recommended Practice for Software Requirements Specifications [6], and The Dept. of Justice Systems Development Life Cycle Guidance Document, Interface Control Document (ICD) Template*, which have been adapted to support ISO 9001 requirements. Additional information is provided in the document template, *Interface Control Document.doc,* which is located on the companion CD-ROM. Table 5-13 provides an example document outline. Additional information in support of associated development work products is provided in Appendix C.

Interface Control Document Guidance

The following provides section-by-section guidance in support of the creation of an ICD. This guidance should be used to help define associated interface control requirements and should reflect the actual requirements of the implementing organization.

*Department of Justice Systems Development Life Cycle Guidance, Interface Control Document Template, Appendix C-17; http://www.usdoj.gov/jmd/irm/lifecycle/table.htm.

Table 5-13. Interface control document outline

Title Page
Revision Page
Table of Contents
1. Introduction
    1.1   System Identification
    1.2   Document Overview
    1.3   References
    1.4   Definitions and Acronyms
2. Description
    2.1   System Overview
    2.2   Interface Overview
    2.3   Functional Allocation
    2.4   Data Transfer
    2.5   Transactions
    2.6   Security and Safety
3. Detailed Interface Requirements
    3.1   Interface 1 Requirements
        3.1.1   Interface Processing Time Requirements
        3.1.2   Message (or File) Requirements
        3.1.3   Communication Methods
        3.1.4   Security Requirements
    3.2   Interface 2 Requirements
4. Qualification Methods

System Identification. This section should contain a full identification of the participating systems, the developing organizations, responsible points of contact, and the interfaces to which this document applies, including, as applicable, identification numbers(s), title(s), abbreviation(s), version number(s), release number(s), or any version descriptors used. A separate paragraph should be included for each system that comprises the interface.

Document Overview. This section should provide an overview of the document, including a description of all sections.

References. This section should list the number, title, revision, and date of all documents referenced or used in the preparation of this document.

Definitions and Acronyms. This section should describe all terms and abbreviations used in support of the development of this document and critical to the comprehension of its content.

Description. A description of the interfaces between the associated systems should be described in the following subsections.

System Overview. This subsection should describe each interface and the data exchanged between the interfaces. Each system should be briefly summarized, with special emphasis on functionality relating to the interface. The hardware and software components of each system should be identified.

Interface Overview. This subsection should describe the functionality and architecture of the interfacing system(s) as they relate to the proposed interface. Briefly summarize each system, placing special emphasis on functionality. Identify all key hardware and software components as they relate to the interface.

Functional Allocation. This subsection should describe the operations that are performed on each system involved in the interface. It should also describe how the end user would interact with the interface being defined. If the end user does not interact directly with the interface being defined, a description of the events that trigger the movement of information should be defined.

Data Transfer. This subsection should describe how data would be moved among all component systems of the interface. Diagrams illustrating the connectivity among the systems are often helpful communication tools in support of this type of information.

Transactions. This subsection should describe the types of transactions that move data among the component systems of the interface being defined. If multiple types of transactions are utilized for different portions of the interface, a separate section may be included for each interface.

Security and Safety. If the interface defined has security and safety requirements, briefly describe how access security will be implemented and how data transmission security and safety requirements will be implemented for the interface being defined.

Detailed Interface Requirements. This section should provide a detailed description of all requirements in support of all interfaces between associated systems. This should include definitions of the content and format of every message or file that may pass between the two systems and the conditions under which each message or file is to be sent. The information presented in subsection Interface 1 Requirements should be replicated as needed to support the description of all interface requirements.

Interface 1 Requirements. Briefly describe the interface, indicating data protocol, communication method(s), and processing priority.

Interface Processing Time Requirements. If the interface requires that data be formatted and communicated as the data is created, as a batch of data is created by operator action, or in accordance with some periodic schedule, indicate processing priority. Priority should be stated as measurable performance requirements defining how quickly data requests must be processed by the interfacing system(s).

Message (or File) Requirements. This subsection should describe the transmission requirements. The definition, characteristics and attributes of the requirements should be described.

Data Assembly Characteristics. This subsection should define all associated data elements that the interfacing entities must provide any required access requirements.

Field/Element Definition. All characteristics of individual data elements should be described.

Communication Methods. This subsection should address all communication requirements to include a description of all connectivity and availability requirements. All aspects of the flow of communication should be described.

Security Requirements. This subsection should address all security features that are required in support of the interface process.

Note: When more than one interface between two systems is being defined in a single ICD, each should be defined separately, including all of the characteristics described in Interface 1 Requirements for each. There is no limit on the number of unique interfaces that can be defined in a single Interface Control Document. In general, all interfaces defined should involve the same two systems.

Qualification Methods. This section should describe all qualification methods to be used to verify that the requirements for the interfaces have been met. Qualification methods may include:

- Demonstration. The operation of interfacing entities that relies on observable functional operation not requiring the use of instrumentation, special test equipment, or subsequent analysis.

- Test. The operation of interfacing entities using instrumentation or special test equipment to collect data for later analysis.

- Analysis. The processing of accumulated data obtained from other qualification methods. Examples are reduction, interpretation, or extrapolation of test results.

- Inspection. The visual examination of interfacing entities, documentation, and so on.

- Special qualification methods. Any special qualification methods for the interfacing entities, such as special tools, techniques, procedures, facilities, and acceptance limits.

If a separate test plan exists, then a reference to this document should be provided.

OPERATION

The operations process contains the activities and tasks of the operator. This process covers the operation of the software product and operational support to users. The operator manages the operation process at the project level following the management process, which is instantiated in this process. It establishes an infrastructure under the process following the infrastructure process; tailors the process for the project following the tailoring process, and manages the process at the organizational level following the improvement process and the training process. Table 5-14 describes the operation process objectives.

Table 5-14. Operation process objectives

---

a) Identify and mitigate operational risks for the software introduction and operation.

b) Operate the software in its intended environment according to documented procedures.

c) Provide operational support by resolving operational problems and handling user inquires and requests.

d) Provide assurance that software (and host system) capacities are adequate to meet user needs.

e) Identify customer support service needs on an ongoing basis.

f) Assess customer satisfaction with both the support services being provided and the product itself on an ongoing basis.

g) Deliver needed customer services.

ISO 9001 Goals

ISO 9001 describes requirements in support of the operator of both software products and services. The specific ISO 9001 Clause, 7.2.3, Customer Communication During Operations and Maintenance, directs the organization to determine and implement effective arrangements for communicating with customers on:

- Software product information
- Inquiries, contracts, or order handling (including amendments)
- Customer feedback (including customer complaints) [47]

Related is Clause 8.2.1, Customer Satisfaction, which requires the monitoring of information on customer perception as to whether the organization is meeting requirements (as one of the performance measurements of the quality management system). The methods for obtaining and using this information must be defined.

The other ISO 9001 Clause, 7.5.1, Control of Production and Service Provision, directs the software-producing organization to plan and control operations, including:

a) The need to set up a help desk to conduct telephone or other electronic communication with the customer(s)

b) Arrangements for ensuring continuity of support, such as disaster recovery, security and backup

IEEE Standard 12207, Standard for Software Life Cycle Processes, supports ISO 9001 product realization. This document guidance is provided because the documentation associated with product delivery and the user's manual is critical in support of the successful transition, or delivery, or any software product. If such user's information is deemed a requirement, the development of user's documentation supports the ISO 9001 quality management requirement for customer communication and also supports requirements associated with customer-related processes.

User's Manual

The following provides section-by-section guidance in support of the creation of a software user's manual or operator's manual. IEEE Std 1063, IEEE Standard for Software User Documentation [18], was the primary reference document for the development of this material.

User's Manual Document Guidance

This guidance should be used to help define an operational process and should reflect the actual processes and procedures of the implementing organization. Additional information is provided in the document template, *Software Users Manual.doc,* which is located on the companion CD-ROM. Table 5-15 provides an example document outline.

Introduction. This section should provide information on document use, all definitions and acronyms, and references.

Document Use. This subsection should describe the intended use of the software user's manual. The organization of the user's document should effectively support its use. If the user's manual is going to contain both instructional and reference material, each type should be clearly separated into different chapters or topics. Task-oriented documentation (instructional) should include procedures that are structured according to user's tasks. Documentation used as reference material should be arranged to provide access to individual units of information. This section can provide an overview of the type of information provided, its intended use, and the organization of the user's manual.

Table 5-15. User's manual document outline

Title Page
Revision Page
Table of Contents
1. Introduction
    1.1   Document Use
    1.2   Definitions and Acronyms
    1.3   References
2. Concept of Operations
3. General Use
4. Procedures and Tutorials
5. Software Commands
6. Navigational Features
7. Error Messages and Problem Resolution
Index

Definitions and Acronyms. This subsection should identify all definitions and acronyms specific to this software user's manual. This should be an alphabetical list of application specific terminology. All terminology used with the user's manual should be consistently applied.

References. This section should provide a list of all references used in support of the development of the software user's manual. Include a listing of all the documentation related to the product that is to be transitioned to the operations area, which includes any security or privacy protection consideration associated with its use. Also include as a part this any licensing information for the product.

Concept of Operations. This section should provide an overview of the software, including its intended use. Descriptions of any relevant business processes or workflow activities should be included. Any items required in support of the understanding of the software product should be included. This may require a description of the theory, method, or algorithm critical to the effective use and understanding of the product.

General Use. Information should be provided in support of routine user activities. It is important to identify actions that will be performed repetitively to avoid redundancy within the user's manual. For example, describing how to cancel or interrupt an operation while using the software would be in this section. Other task-oriented routine documentation could include software installation and deinstallation procedures, how to log on and off the application, and the identification of basic items/actions that are common across the applications' user interface.

Procedures and Tutorials. Information of a tutorial (i.e., procedural) nature should be provided in the user's documentation as clearly as possible. A consistent approach to the presentation of the material is important when trying to clearly communicate a concept to the user.

Describe the purpose and concept for the tutorial information presented in the user's manual. Include a list of all activities that must be completed prior to the initiation of the procedure or tutorial. Identify any material that should be used as reference in support of the task. List all cautions and other supporting information that are relevant in supporting the performance of the task.

It is important to list all instructional steps in the order in which they should be performed, with any optional steps clearly identified. The steps should be consecutively numbered and the initial and last steps of the task should be clearly identified. It is important that the user understand how to successfully initiate and complete the procedure or tutorial.

Warnings and cautions should be distinguishable from instructional steps and should be preceded by a word and graphic symbol alerting the user to the item. For example, *warning (graphic)* would precede a warning to the user. The use of the following format for warning and cautions is suggested: *word and graphic, brief description, instructional text, description of consequences,* and *proposed solution or workaround.*

Software Commands. The user's manual should describe all software commands, including required and optional parameters, defaults, precedence, and syntax. All reserved words and commands should be listed. This section should not only provide the commands, but should also provide examples of their use. Documentation should include a visual representation of the element, a description of its purpose, and an explanation of its intended action. A quick-reference card may be included in the user's documentation to provide the user with the ability to rapidly refer to commonly used commands.

Navigational Features. The document should describe all methods of navigation related to the software application. All function keys, graphical-user-interface items, and commands used in support of application navigation should be described and supported with examples.

Error Messages and Problem Resolution. Information in support of problem resolution (i.e., references) should address all known problems or error codes present in the software application. Users should be provided information that will either help them recover from known problems or report unknown issues, and suggest application enhancements.

Index. An index provides an effective way for users to access documented information. It is important to remember that for an index to be useful it should contain words that users are most likely to look up and should list all topics in the user's documentation. Pay special attention to the granularity and presentation of the index topics. Place minor key words under major ones; for example, instead of using *files* with 30 pages listed, use files, saving and *files, deleting,* with their associated specific pages listed.

MAINTENANCE

The maintenance process contains the activities and tasks of the maintainer. The objective is to modify the existing software product while preserving its integrity. The maintainer manages the maintenance process at the project level following the management process, which is instantiated in this process; establishes an infrastructure under the process following the infrastructure process; tailors the process for the project following the tailoring process; and manages the process at the organizational level following the improvement process and the training process. Table 5-16 describes the maintenance process objectives.

This process consists of the following activities:

1. Process implementation
2. Problem and modification analysis
3. Modification implementation
4. Maintenance review/acceptance
5. Migration
6. Software retirement [47]

Table 5-16. Operation process objectives

a) Define the impact of organization, operations, and interfaces on the existing system in operation.
b) Identify and update appropriate life cycle data.
c) Develop modified system components with associated documentation and tests that demonstrate that the system requirements are not compromised.
d) Migrate system and software upgrades to the user's environment.
e) Ensure that fielding of new systems or versions does not adversely affect ongoing operations.
f) Maintain the capability to resume processing with prior versions.

ISO 9001 Goals

ISO 9001 describes requirements in support of the maintainer of both software products and services. The key ISO 9001 Clause, 7.3.7, Control of Design and Development Changes, directs the maintainer to identify and manage records on all design and development changes. These changes must be reviewed, verified, validated, and approved before implementation. Changes must be evaluated in terms of their effect on constituent parts and products already delivered. The results of the change review and subsequent follow-up actions must be maintained.

Another requirement is ISO 9001 Clause 7.2.3, Customer Communication during Operations and Maintenance, which directs the maintainer to determine and implement effective arrangements for communicating with customers on:

- Software product information
- Inquiries, contracts, or order handling (including amendments)
- Customer feedback (including customer complaints) [47]

Finally, ISO 9001 Clause 7.5.1, Control of Production and Service Provision, directs the maintainer to plan and control operations, including arrangements for ensuring continuity of support, such as disaster recovery, security, and backup.

Transition Plan

The purpose of transition planning is to lay out the tasks and activities that need to take place to efficiently move a product (i.e., specify product name, in-house developed software or COTS software, middleware or component software/hardware) from the development or pilot environment to the production, operations, and maintenance environment. The transition planning steps apply whether the product is being transitioned within an agency (i.e. from agency development staff to agency network or operations staff) or to an outside agency (i.e., information technology systems).

The transition plan is designed to facilitate migration of an application system from development to production (i.e., maintenance). The information provided here in support of transition plan development is based upon IEEE Std-1219, IEEE Standard for Software Maintenance [22], IEEE Std 12207.0, Standard for Information Technology—Life Cycle Processes [39], and US DOD Data Item Description DI-IPSC-81429, Software Transition Plan [71] as primary reference material. Additional information has been incorporated as "lessons learned" from multiple production-application systems and transition opportunities. Table 5-17 provides a proposed document outline in support of ISO9001 requirements.

Table 5-17. Transition plan document outline

Title Page
Revision Page
Table of Contents
  1. Introduction
      1.1   Overview
      1.2   Scope
      1.3   Definitions and Acronyms
            1.3.1   Key Acronyms
            1.3.2   Key Terms
      1.4   References
  2. Product
      2.1   Relationships
  3. Strategies
      3.1   Identify Strategy
      3.2   Select Strategy
  4. Transition Schedules, Tasks, and Activities
      4.1   Installation
      4.2   Operations and Support
      4.3   Conversion
      4.4   Maintenance
  5. Resource Requirements
      5.1   Software Resources
      5.2   Hardware Resources
      5.3   Facilities
      5.4   Personnel
      5.5   Other Resources
  6. Acceptance Criteria
  7. Management Controls
  8. Reporting Procedures
  9. Risks and Contingencies
  10. Transition Team Information
  11. Transition Impact Statement
  12. Plan Review Process
  13. Configuration Control

Transition Plan Document Guidance

The following provides section-by-section guidance in support of the creation of a software transition plan. The development of a software transition plan is critical to the successful transition of software from development to deployment. This guidance should be used to help develop and define the transition process and should reflect the actual processes and procedures of the implementing organization. Additional information is provided in the document template, *Software Transition Plan.doc,* which is located on the companion CD-ROM.

> *Overview.* The transition plan should include an introduction addressing background information on the project. Show the relationship of the project to other projects and/or organizations or agencies, address maintenance resources required, identify the transition team's organization and responsibilities, as well as the tools, techniques, and methodologies that are needed to perform an efficient and effective transition.
>
> The transition plan should include deployment schedules, resource estimates, identification of special resources, and staffing. The transition plan shall also define management controls and reporting procedures, as well as the risks and contingencies. Special attention must be given to minimizing operational risks. An impact statement should be produced outlining the potential impact of the transition to the existing infrastructure, operations, and support staff, and to the user community.
>
> *Scope.* This subsection should include a statement of the scope of the transition plan. Include a full identification of the product to which this document applies including (as applicable) identification numbers, titles, abbreviations, version numbers, and release numbers. Include:
>
> > Product overview
> >
> > Overview of the supporting documentation
> >
> > Description of the relationship of the product to other related projects and agencies
>
> *Definitions and Acronyms.* This subsection should identify all definitions and acronyms specific to this software configuration management plan.
>
> *References.* This subsection should provide a list of all references used in support of the development of the software configuration management plan. Include a listing of all the documentation related to the product that is to be transitioned to the operations area, including any security or privacy protection considerations associated with its use. Include as a part of this any licensing information for the product.
>
> *Product.* This section should include a brief statement of the purpose of the product to which this document applies. It should also describe the general nature of the product; summarize the history of development, operation, and maintenance; identify the project sponsor, acquirer, user, developer, vendor,

and maintenance organizations; identify current and planned operating sites; and list other relevant documents.

Relationships. This subsection should describe the relationship(s) of the product being transitioned to any other projects and agencies. The inclusion of a diagram or flow chart to help indicate these relationships is often helpful.

Identify Strategies. This subsection should identify the transition strategies and tools to be used as part of the transition plan. Identify all the options for moving the product from its present state into production/operations. These options could include:

Incremental implementation or phased approach

Parallel execution

One-time conversion and switchover

Any combination of the above

Each option should also identify the advantages and disadvantages, risks, estimated time frames, and estimated resources.

Select Strategy. This subsection should include an evaluation of each of the transition options, comparing them to the transition requirements, and selecting the one that is most appropriate for the project. Once a transition strategy has been selected, then the justification is documented and approved.

Transition Schedules, Tasks, and Activities. This section should include, or provide reference to, detailed schedules for the selected transition strategy. These schedules should include equipment installation, training, conversion, deployment, and or retirement of the existing system (if applicable), as well as any transition activities required to turn over the product from developers or vendors to operational staff. The schedules should reflect all milestones for conducting transition activities.

Also include an installation schedule for equipment (new or existing), software, databases, and so on. Include provisions for training personnel with the operational software and target computer(s), as well as any maintenance software and/or host system(s). Describe the developer's plans for transitioning the deliverable product to the maintenance organization. This should address the following:

Planning/coordination of meetings

Preparation of items to be delivered to the maintenance organization

Packaging, shipment, installation, and checkout of the product maintenance environment

Packaging, shipment, installation, and checkout of the operational software

Training of maintenance/operational personnel

Installation. Installation consists of the transportation and installation of the product from the development environment to the target environment(s). This subsection should describe any required modifications to the product, checkout in the target environment(s), and customer acceptance procedures. If problems arise, these should be identified and reported, these procedures should be addressed here as well. If known, any temporary "work-around(s)" should also be described.

Operations and Support. This subsection should address user operations and all required ongoing support activity. Support includes providing technical assistance, consulting with the user, and recording user support requests by maintaining a support request log. The operations and support activities can trigger maintenance activities via the ongoing project monitoring and controlling activities or problem and change logs, and this process should be described or a reference to associated documentation should be provided.

Conversion. This subsection should address any data or database transfers to the product and its underlying components that would occur during the transition.

Maintenance. Maintenance activities are concerned with the identification of enhancements and the resolution of product errors, faults, and failures. The requirements for software maintenance initiate "service-level changes" or "product modification requests" by using defined problem and change management reporting procedures. This section should describe all issues and activities associated with product maintenance during the transition.

Resource Requirements. All estimates for resources (hardware, software, and facility) as well as any special resources (i.e., service and maintenance contracts) and staffing for the selected transition strategy should be described in this section. The assignment of staff, agency, and vendor responsibility for each task identified should be documented. This allows managers and project team members to plan and coordinate the work of this project with other assignments. If specific individuals cannot be identified when the transition plan is developed, generic names may be used and replaced with individual names as soon as the resources are identified.

Software Resources. A description of any software and associated documentation needed to maintain the deliverable product should be included in this subsection. The description should include specific names, identification numbers, version numbers, release numbers and configurations as applicable. References to user/operator manuals or instructions for each item should be included. Identify for each product item where it is to come from—acquirer-furnished, currently owned by the organization, or to be purchased. Include information about vendor support, licensing, and usage and ownership rights, whether the item is currently supported by the vendor, whether it is expected to be supported at the time of delivery, whether licenses will be assigned to the maintenance organization, and the terms of such licenses. Include any required service and maintenance contract costs as well as payment responsibility.

Hardware Resources. This subsection should include a description of all hardware and associated documentation needed to maintain the deliverable product. This hardware may include computers, peripheral equipment, simulators, emulators, diagnostic equipment, and noncomputer equipment. The description should include specific models, versions, and configurations. References to user/operator manuals or instructions for each item should be included. Identify each hardware item and document as acquirer-furnished, an item that will be delivered to the maintenance organization, or an item the maintenance organization currently owns or needs to acquire. (If the item is to be acquired, include information about a current source of supply, and order information as well as what budget is to pay for it.) Include information about manufacturer support, licensing, and usage and ownership rights, whether the items are currently supported by the manufacturer, or will be in the future, and whether licenses will be assigned to the maintenance organization and the terms of such licenses.

Facilities. Describe any facilities needed to maintain the deliverable product in this subsection. These facilities may include special buildings, rooms, mockups, building features such as raised flooring or cabling, building features to support security and privacy protection requirements, building features to support safety requirements, special power requirements, and so on. Include any diagrams that may be applicable.

Personnel. This subsection should include a description of all personnel needed to maintain the deliverable product, including anticipated number of personnel, types of support personnel (job descriptions), skill levels and expertise requirements, and security clearance.

Other Resources. Identify any other consumables (i.e., technology, supplies, and materials) required to support the product. Provide the names, identification numbers, version numbers, and release numbers. Identify if the document or consumable is acquirer-furnished, an item that will be delivered to the maintenance organization, or an item the organization current owns or needs to acquire. If the maintenance/operational organization needs to acquire it, identify what budget will cover the expense.

Acceptance Criteria. It is important to establish the exit or acceptance criteria for transitioning the product. These criteria will determine the acceptability of the deliverable work products and should be specified in this section. Representatives of the transitioning organization and the acquiring organization should sign a formal agreement, such as a service-level agreement, that outlines the acceptance criteria. Any technical processes methods or tools, as well as performance benchmarks required for product acceptance, should be specified in the agreement. Also include an estimation of the operational budget for the product and how these expenses will be covered.

Management Controls. Describe all management controls to ensure that each task is successfully executed and completed based on the approved acceptance criteria. This should include procedures for progress control, quality control, change control, version control, and issue management during the transition process.

Reporting Procedures. This section should define the reporting procedures for the transition period. Such things as type of evaluations (review, audit, or test) as well as anomalies that are identified during the performance of these evaluations should be reported.

Risks and Contingencies. This section should identify all known risks and contingencies faced by the transition process, with special attention given to minimizing operational risks. Reference to an organization risk management plan should be provided here. A description of risk mitigation, tracking, and reporting should be presented.

Transition Team Information. This section should include all transition team information, including the transition team's organization, roles and responsibilities for each activity, as well as the tools, techniques, and methodologies and/or procedures that are needed to perform the transition.

Transition Impact Statement. This section should contain a transition impact statement that describes any anticipated impact to existing network infrastructure, support staff, and user community during the system transition. The impact statement should include descriptions for the performance requirements, availability, security requirements, expected response times, system backups, expected transaction rates, initial storage requirements with expected growth rate, as well as help-desk support requirements.

Plan Review Process. This section should describe the review procedure/process in support of this document. A review should be held to identify and remove any defects from the transition plan before it is distributed. This is a content review and should be conducted by the appropriate members of the project team or an independent third party. The results should be recorded in anInspection Report and Inspection Log Defect Summary, shown in Appendix C.

Configuration Control. The transition plan information should be subject to the configuration control process for the project. Subsequent changes are tracked to ensure that the configuration of the transition plan information is known at all times. Changes should be allowed only with the approval of the responsible authority. Transition plan changes should follow the same criteria established for the project's change control procedures and reference to any associated project level configuration management plan should be provided here.

Chapter 6

12207 Supporting Life Cycle Processes and ISO 9001

SUPPORTING PROCESSES

IEEE 12207 describes eight supporting processes that support the project from the organizational level:

1. Documentation
2. Configuration management
3. Quality assurance
4. Verification
5. Validation
6. Joint review
7. Audit
8. Problem resolution

As described by ISO 9001, the organization employing and performing a supporting process:

- Manages it at the project level following the management process.
- Establishes an infrastructure under it following the infrastructure process.
- Tailors it for the project following the tailoring process.
- Manages it at the organizational level following the improvement process and the training process.
- Assures its quality through joint reviews, audits, verification, and validation.

ISO 9001 Goals

ISO 9001 describes requirements in support of the IEEE 12207 supporting processes through almost 30 requirements for the organization that performs these eight supporting processes. Dominant are two clauses. First, ISO 9001 Clause 4.2, Documentation requirements, results in documented processes and records established and maintained as evidence of conformance to requirements. Second, as the supporting processes are focused on product realization, Clause 7.1, Planning of Product Realization, results in project documents and records. For each supporting process below, the applicable ISO 9001 requirement clauses will be detailed. Figure 6.1 shows the context of supporting processes within the SLC processes.

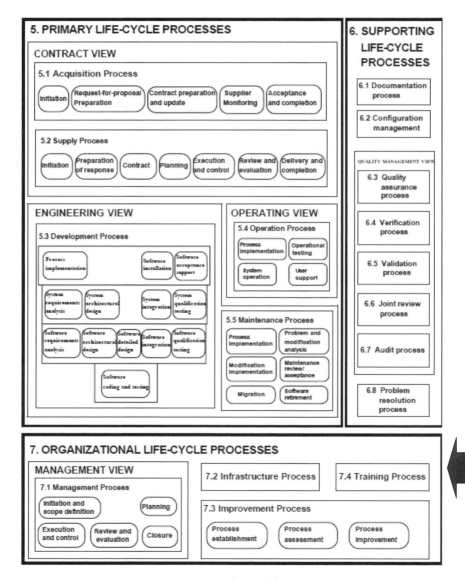

Figure 6.1.

DOCUMENTATION

The documentation process is a process for recording information produced by a life cycle process or activity. This process contains the set of activities to plan, design, develop, produce, edit, distribute, and maintain those documents needed by all concerned, such as managers, engineers, and users of the software product. Table 6-1 provides a list of the document process objectives.

ISO 9001 Goals

ISO 9001 describes requirements directly in support of the IEEE 12207 documentation process. Several standard clauses are stipulated. Since the documentation process is an organizational responsibility rather than a project responsibility, both ISO 9001 Clauses 4 and 7 apply. Clause 4.2.1, General Documentation Requirements, for the management system documentation, includes:

- Documented statements of policies and related objectives
- Documents needed by the organization for effective planning, operation, and control of its processes
- A manual that links together all document and records types
- Records showing conformance to policies and plans

This general clause covers the following three detailed clauses: Quality Manual or management system overview and index, control of documents, and records. Clause 4.2.2, Quality Manual, requires a manual that links together all management system documents and records types. This includes management system scope, justification for any exclusion, references to all procedures, and description of interaction between processes. Clause 4.2.3, Control of Documents, requires management system documents to be controlled. This includes document identification, review, approval, updates, change status, distribution availability, and retirement. A similar Clause, 4.2.4, Control of Records, requires records to be established and maintained as evidence of conformance to requirements and to demonstrate the effective operation of the management system. Records must remain legible, readily identifiable, and retrievable.

Table 6-1. Document process objectives

a) Identify all documents to be produced by the process or project.
b) Specify the content and purpose of all documents and plan and schedule their production.
c) Identify the standards to be applied for development of documents.
d) Develop and publish all documents in accordance with identified standards and in accordance with nominated plans.
e) Maintain all documents in accordance with specified criteria.

Next is Clause 7.1, Planning of Product Realization, for the planning and development of processes. The planning must determine the:

- Quality objectives and software product requirements
- Need for processes, documents, and resources
- Verification, validation, monitoring, inspection, and test activities
- Criteria for software product acceptance
- Records as evidence that the processes and resulting software product meet requirements

The documents designed should be in a form suitable for the organization. Note that this book contains over 100 document outlines with individual field descriptions.

Lastly, a similar requirement to Clause 7.1 is Clause 7.3.7, Control of Design and Development Changes, as design and development changes must be identified and records maintained. Although many of the "new software product" process documents can be reused for changes, record keeping is critical, as the changes must be reviewed, verified, and validated, and approved before implementation. The results of the change review and subsequent follow-up actions must be maintained.

Quality Manual

The following is provided as an example table of contents supporting the development of a quality manual for a software or systems development effort.

1. INTRODUCTION
    1.1   Purpose
    1.2   Background
    1.3   Scope and Applicability
2. QUALITY SYSTEM FRAMEWORK
    2.1   Quality Policy
    2.2   Methodology
    2.3   Quality Roles and Responsibilities
    2.4   Quality System Reviews
    2.5   Methodology Maintenance
    2.6   Quality Manual Maintenance
3. LIFE CYCLE ACTIVITIES
    3.1   Proposal Preparation
    3.2   Contracting
    3.3   Project Initiation
        3.3.1   Peer Review
        3.3.2   Task Estimating
        3.3.3   Project Plan
        3.3.4   Quality Assurance Plan
        3.3.5   Staffing

Configuration Management Record

Many existing software configuration management systems can be expanded to plan and track all organizational and project documents and related records. An example of a configuration management record is shown in Table 6-2.

CONFIGURATION MANAGEMENT

The configuration management process works throughout the software life cycle to:

- Identify, define, and baseline software configuration items (CI) in a system.
- Control modifications and releases of the CI (e.g., updating of multiple products in one or more locations).

Table 6-2. Configuration management record

Configuration item (CI) identifier
CI Type (e.g., policy, procedures, plan, project artifact, report, record)
CI Entity control level (e.g., organization, project, system, subsystem, customer)
CI Full title
CI Version
CI Status (e.g., V&V, CR#)
CI Owner name
CI Owner contact information
CI Date, last updated
CI References
CI Location
CI List of keywords
CI Links to other information sources
Change request (CR) identifier
CR Requester name
CR Requester contact information
CR Date
CR Title
CR Description
CR Justification
CR Suggested effective date
CR Acceptance disposition
CR Implementer information
CR Targeted effective date
CR Brief discussion
CR Date of last data entry
CR Detailed summary of the results, including objectives and procedures
CR Validation tracking (e.g., test phases)
CR Validation contact information

- Record and report the status of the CI and modification requests (e.g., including all actions and changes resulting from a change request or problem).
- Ensure the completeness, consistency, and correctness of the CI (e.g., verification and validation activities).
- Control storage, handling, and delivery of the CI (e.g., release management and delivery).

The CI is defined within a configuration that satisfies an end use function and can be uniquely identified at a given reference point. Table 6-3 provides a list of the configuration management process objectives.

ISO 9001 Goals

ISO 9001 describes requirements in support of the IEEE 12207 configuration management process, starting with Clause 7.5.3, Identification and Traceability. This key clause requires software product identification during product realization and the software product status to be identified with testing requirements. The other ISO 9001 requirements follow:

Clause 7.1, Planning of Product Realization, requires planned product realization processes, including software configuration management plans (SCMPs).

Clause 7.3.1, Design and Development Planning, requires planning to identify software configuration management (SCM) practices.

Clause 7.3.7, Control of Design and Development Changes, requires changes to be identified, reviewed, verified and validated, evaluated, and approved before implementation. The control of the design and development changes is part of SCM process, which maintains the records.

Clause 7.5.1, Control of Production and Service Provision, requires the implementation of release, delivery, and postdelivery activities used in the build, release, and replication of the software configuration items. Depending on the contract, software and hardware configuration maybe defined for a specific installation.

Clause 7.5.4, Customer Property, requires that care be taken with any customer property while it is under the control of, or being used by, the organization.

Table 6-3. Configuration management process objectives

a) Identify, define, and control all relevant items of the project
b) Control modifications of the items
c) Record and report the status of items and modification requests
d) Ensure the completeness of the items
e) Control storage, handling, release, and delivery of the items

Any lost, damaged, or unsuitable property must be recorded and reported to the customer.

Clause 7.6, Control of Measuring and Monitoring Device, requires test planning and execution for the monitoring and measuring to provide evidence of software product conformance to requirements. The SCM system provides the control of testing devices, where testing devices include:

    a) data used for testing the software product

    b) software tools

    c) computer hardware

    d) instrumentation interfacing to the computer hardware

Clause 8.3, Control of Nonconforming Product or Erroneous Configuration Item (CI), requires the nonconforming CI be identified and controlled to prevent its unintended use or delivery. The configuration management system provides the controls and records of nonconforming and revised CIs and the associated versions.

Clause 8.5.2, Corrective Action, requires action to eliminate the cause of the nonconformity and prevent recurrence. Configuration management can manage the changes when corrective action directly affects the software products.

Software Configuration Management Plan

ISO 9001 fully supports the requirement to establish and maintain a plan for performing all configuration management process activities. ISO 9001 requires the documentation of project-level CM activities, but also requires the description of organizational CM activities. IEEE Std 828, IEEE Standard for Software Configuration Management Plans (SCMP), can be used to help support this requirement. Appendix C also provides examples of supporting CM work products, which include a Configuration Control Board (CCB) Letter of Authorization, CCB Charter, and Software Change Request Procedures. The modification of the recommended CM table of contents to support the goals of ISO 9001 more directly is shown in Table 6-4.

Software Configuration Management Plan Document Guidance

The following provides section-by-section guidance in support of the creation of a software configuration management plan (SCMP). The SCMP should be considered to be a living document and should change to reflect any process improvement activity. This guidance should be used to help define a software configuration management process and should reflect the actual processes and procedures of the implementing organization. Additional information is provided in the document template, *Software Configuration Management Plan.doc,* which is located on the companion CD-ROM.

Table 6-4. Software configuration management plan document outline

Title Page
Revision Page
Table of Contents
1.0 Introduction
    1.1 Purpose
    1.2 Scope
    1.3 Definitions/Acronyms
    1.4 References
2.0 Software Configuration Management
    2.1 SCM Organization
    2.2 SCM Responsibilities
    2.3 Relationship of SCM to the Software Process Life Cycle
        2.3.1 Interfaces to Other Organizations on the Project
        2.3.2 Other Project Organizations SCM Responsibilities
    2.4 SCM Resources
3.0 Software Configuration Management Activities
    3.1 Configuration Identification
        3.1.1 Specification Identification
        3.1.2 Change Control Form Identification
        3.1.3 Project Baselines
        3.1.4 Library
    3.2 Configuration Control
        3.2.1 Procedures for Changing Baselines
        3.2.2 Procedures for Processing Change Requests and Approvals
        3.2.3 Organizations Assigned Responsibilities for Change Control
        3.2.4 Change Control Boards (CCBs)
        3.2.5 Interfaces
        3.2.6 Level of Control
        3.2.7 Document Revisions
        3.2.8 Automated Tools Used to Perform Change Control
    3.3 Configuration Status Accounting
        3.3.1 Storage, Handling and Release Of Project Media
        3.3.2 Information and Control
        3.3.3 Reporting
        3.3.4 Release Process
        3.3.5 Document Status Accounting
        3.3.6 Change Management Status Accounting
    3.4 Configuration Audits and Reviews
4.0 Configuration Management Milestones
5.0 Training
6.0 Subcontractor Vendor Support

Introduction

The introduction information provides a simplified overview of the software configuration management (SCM) activities so that those approving, those performing, and those interacting with SCM can obtain a clear understanding of the plan. The introduction should include four topics: the purpose of the plan, the scope, the definition of key terms, and references.

Purpose

The purpose should address why the plan exists and its intended audience.

Scope

The scope should address SCM applicability, limitations, and assumptions on which the plan is based. The scope should provide an overview description of the software development project, identification of the software CI(s) to which SCM will be applied, identification of other software to be included as part of the plan (e.g., support or test software), and relationship of SCM to the hardware or system configuration management activities for the project. The scope should address the degree of formality, depth of control, and portion of the software life cycle for applying SCM on this project, including any limitations, such as time constraints, that apply to the plan. Any assumptions that might have an impact on the cost, schedule, or ability to perform defined SCM activities (e.g., assumptions of the degree of customer participation in SCM activities or the availability of automated aids) must also be addressed. The following is an example of scope.

> *This document defines SCM activities for all Software and Data produced during the development of the [Project Name] ([Project Abbreviation]) software. This document applies to all module products, end-user products, and data developed and maintained for the [Project Abbreviation] Program. CM activities as defined herein will be applied to all future [Project Abbreviation] projects.*
>
> *This document conforms to [Company Name]'s Software Configuration Management Policy [Policy #] and IEEE standards for software configuration management, and will change as needed to maintain conformance.*

Software Configuration Management (SCM)

Appropriately documented SCM information describes the allocation of responsibilities and authorities for SCM activities to organizations and individuals within the project structure. SCM management information should include three topics: the project organization(s) within which SCM is to apply, the SCM responsibilities of these organizations, and references to the SCM policies and directives that apply to this project.

SCM Organization

The SCM organizational context must be described. The plan should identify the all participants, or those responsible for any SCM activity on the project. All functional roles should be described, as well as any relationships to external organizations. Organization charts, supplemented by statements of function and relationships, can be an effective way of presenting this information. An example is shown in Figure 6-2.

SCM Responsibilities

All those responsible for SCM implementation and performance should be described in this section. A matrix describing SCM functions, activities, and tasks can be useful for documenting the SCM responsibilities. For any review board or special organization established for performing SCM activities on this project, the plan should describe its

a) Purpose and objectives

b) Membership and affiliations

c) Period of effectivity

d) Scope of authority

e) Operational procedures

The following provides an example of typical responsibilities.

The Program Manager (PM) is responsible for ensuring the SCM process is developed, maintained and implemented. The PM is responsible for ensuring that project leads and team members are adequately trained in SCM policy and procedure. The [Project Abbreviation] Program Manag-

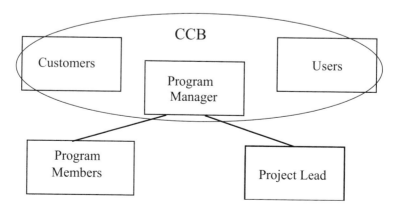

Figure 6-2. Example SCM organization.

er and Project Leads work together to develop, maintain, and implement effective software configuration management. Please refer to Figure 3.1 and 3.1.

The [Project Abbreviation] Program Manager (PM) has the authority to ensure development of an appropriate configuration management process. Since CCB responsibility exists with the customer (See Figure 3.1), team member(s) of the [Project Abbreviation] Project will be identified as members of this CCB and will represent our interests there.

The Configuration Control Board (CCB) is responsible for the review and approval of all Software Change Requests (SCRs), all baseline items, and all changes to baseline items. The CCB considers the cost and impact of all proposed changes, and considers the impact to all interfacing items as part of their approval action. The CCB for [Project Abbreviation] consists of the PM, the PL, and two or more members from the government Program Management team.

The associated Configuration Control Board (CCB) has the authority to approve a baseline version of this process, as well as any changes to the process as recommended by any associated CCB representative. The same CCB has the authority to define the baseline for each software module, and end-user products, and to approve start of work and acceptance of completed changes to the baseline products. This CCB also has the authority to approve reversion to a previous version of the product if warranted.

The [Project Abbreviation] Program SCM is responsible for implementing the actions of this process and for the implementation of [Company Name] process improvement recommendations.

The Software Configuration Manager ensures that the process is implemented once approved and that proposed changes to the process are reviewed by [Project Abbreviation] team members and approved by the CCB as required.

Project Leads, or a designee, are responsible for adhering to the [Project Abbreviation] SCM process and for ensuring that only approved changes are entered into the module product baselines. Project Leads, or a designee, are also responsible for identifying the module version to be included in the end-user products. The Project Lead, or a designee, is responsible for controlling the software product(s) development and/or maintenance baseline to ensure that only approved changes are entered into the baseline(s) system. The Project Lead maintains the software product with all changes to be approved by the associated CCB.

The Project Lead (PL) develops and proposes a baseline version of the software configuration management product, ensuring that the SCM process is followed, once approved, and proposes process changes to the [Project Abbreviation] SCM manager and CCB, as required.

Each Project Lead (PL) is responsible for ensuring that project configuration management processes are followed. The PL also ensures all members of the project team are trained in SCM policies and procedures that apply to their function on the team. This consists of initial training for new members and refresher training, as required, due to process or personnel changes.

Relationship of SCM to the Software Process Life Cycle

This section should describe SCM as it relates to other elements of the software process life cycle. A diagram showing the relationship between SCM elements and the overall process is often used here.

SCM Resources

This section should describe the minimum recommended resources (e.g., time, tools, and materials) dedicated to software configuration management activities.

Configuration Identification

All technical and managerial SCM activities should be described, as well as all functions and tasks required to manage the configuration of the software. General project activities that have SCM implications should also be described from the SCM perspective. SCM activities are traditionally grouped into four functions: configuration identification, configuration control, status accounting, and configuration audits and reviews. An example of the type of information provided in this section follows.

> *All [Project Abbreviation] Modules and baselines are to be approved by the CCB. Once a module has been baselined all changes made will be identified in their respective Version Description Documents (VDDs).*
>
> *The [Project Abbreviation] is currently in development, product baseline to be approved upon delivery.*
>
> *Product baselines are to be maintained by the [Project Abbreviation] SCM. The Baseline will consist of the following:*

- *Source files*
- *Libraries*
- *Executable files*
- *Data files*
- *Makefiles*
- *Link files*
- *VDDs*
- *User's manuals*

Specification Identification

SCM activities should identify, name, and describe all code, specifications, design, and data elements to be controlled for the project. Controlled items may include executable code, source code, user documentation, program listings, databases, test cases, test plans, specifications, management plans, and elements of the support environment (such as compilers, operating systems, programming tools, and test beds).

Change Control Form Identification

The plan should specify the procedures for requesting a change. As a minimum, the information recorded for a proposed change should contain the following:

a) The name(s) and version(s) of the CIs where the problem appears
b) Originator's name and organization
c) Date of request
d) Indication of urgency
e) The need for the change
f) Description of the requested change

Additional information, such as priority or classification, may be included to assist in its analysis and evaluation. Other information, such as change request number, status, and disposition, may be recorded for change tracking. An example of the type of information provided in this section follows:

> *Software Change Request (SCR) procedures for each development team are defined in Appendix A. These procedures are used to add to, change, or remove items from the baselines. The identification and tracking of change requests is accomplished through Change Enhancement Requests (CER).*

Project Baselines

This section should describe the configuration control activities required implement changes to baselined CIs. The plan should define the following sequence of specifc steps:

a) Identification and documentation of the need for a change
b) Analysis and evaluation of a change request
c) Approval or disapproval of a request
d) Verification, implementation, and release of a change

The plan should identify the records to be used for tracking and documenting this sequence of steps for each change.

Library

This section should identify the software configuration management repository, if one exists. The following provides an example of the type of typical information:

[Project Abbreviation] is composed of end-user products, module products and data. The organization of these computer software units is given in Appendix B.

Procedures for Changing Baseline

This section should describe the procedures for making controlled changes to the development baseline. The following provides an example of this type of information.

Changes to each software product baseline are made according to the change process.

Change Tracking. Changes to each software product baseline are tracked using an automated CER and SCR tracking system. These systems report current status for all CERs and SCRs for a specified version and product.

Change Release. The PL provides reports containing the following information for review at each CCB meeting for:

a) Number of new CERs
b) Type
c) Priority
d) Source (internal, field, FOT&E)
e) Action required
f) Number of CERs in each status

Each development team Project Lead, or designated Change Request Coordinator (CRC), is responsible for keeping the information in the CER tracking system up to date so the information is available to the PL and PM prior to CCB meetings.

Procedures for Processing Change Requests and Approvals

This section should describe the procedures for processing changes requests and the associated approval process. An example of a change request process is provided in Appendix C, Software Change Request Procedures, of this text.

Organizations Assigned Responsibilities for Change Control

This section should describe all organizations associated with the change control process and their associated responsibilities.

Change Control Boards (CCBs)

This section should describe the role of the CCB and refer to any associated documentation (e.g., CCB Charter).

Interfaces

This section should describe any required software interfaces and point to the source of their SCM documentation.

Level of Control

This section should describe the levels of control and implementing authorities. The following is an example:

> Evaluating Changes. Changes to existing software baselines are evaluated by the software product's associated CCB.
>
> Approving or Disapproving Changes. Changes to existing software baselines are either approved or disapproved by the associated project CCB.
>
> Implementing Changes. Approved changes are incorporated into the product baseline by the development team as described by the change control process documented in Appendix A.

Document Revisions

This section should describe SCM plan maintenance information . All activities and responsibilities necessary to ensure continued SCM planning during the life cycle of the project should be described. The plan should describe:

a) Who is responsible for monitoring the plan

b) How frequently updates are to be performed

c) How changes to the plan are to be evaluated and approved

d) How changes to the plan are to be made and communicated

Automated Tools Used to Perform Change Control

This section should identify all software tools, techniques, equipment, personnel, and training necessary for the implementation of SCM activities. SCM can be performed by a combination of software tools and manual procedures. For each software tool, whether developed within the project or brought in from outside the project, the plan should describe or reference its functions and shall identify all configuration controls to be placed on the tool.

Configuration Status Accounting

Configuration status accounting activities record and report the status of a project's CIs. The SCM plan should describe what will be tracked and reported, describe the types of reporting and its frequency, and how the information will be processed and controlled.

If an automated system is used for any status accounting activity, its function should be described or referenced. The following minimum data elements shall be tracked and reported for each CI: its initial approved version, the status of requested changes, and the implementation status of approved changes. The level of detail and specific data required may vary according to the information needs of the project and the customer.

Storage, Handling, and Release of Project Media

This section should describe the storage, handling and release of SCM products. The following is provided as an example:

> *The initial approved version of a [Project Abbreviation] software product is separately maintained by the PL until the approval of subsequent version of the product. Upon approval of the subsequent version the initial version is held in the project SCM library. This library is under configuration control.*

Reporting

This section should describe the format, frequency, and process for reporting SCM product baseline status. Providing configuration items for testing may require a test item transmittal report. This test item report should identify the person responsible for each item, the physical location of the item, and item status. The following is an example of status report of requested changes:

> *The status report of requested changes is available to the PM at any time. A schedule status report may be provided to the CCB weekly and reported to [Company Name] senior management on a monthly basis. Please refer to the [Project Abbreviation] Software Development Plan, Section 6, for a status report schedule example.*

Release Process

This section should describe the release process, including all required approvals. A summary chart is sometimes used to describe the release process effectively.

Document Status Accounting

This section should describe how changes to the SCM plan are managed and reported.

Change Management Status Accounting

This section should describe the format, frequency, and process for reporting SCM change management status.

Configuration Audits and Reviews

This section should provide information in support of required configuration audits and reviews. The following is provided as an example.

> *Each team PL reviews the software baseline semi-annually. The [Project Abbreviation] PM and [Project Abbreviation] SCM review the baseline of each software product annually. This baseline review consists of checking for CER implementation. The CERs are randomly selected. Discrepancies will be annotated and reported to the PL. Corrective action is taken if required.*
>
> *The [Project Abbreviation] CCB reviews the SCM process annually. Randomly selected problem reports are identified and walked-through the change process. Discrepancies are annotated and reported to the [Project Abbreviation] PM and [Project Abbreviation] SCM. Corrective action is taken if required.*

A software baseline, the associated status reporting, and associated documentation for each software product is available for SQA review at all times.

SCM Milestones

This section should provide a description of the minimum set of SCM-related project milestones that are acceptable for compliance. These milestones should be reflected in the software project management plan schedule and resource allocation.

Training

This section can refer to an independent training plan. However, this training plan must include information relating specifically to SCM training. If an independent plan does not exist, then information regarding the type and frequency of SCM training should be identified here.

Subcontractor/Vendor Support

This section should describe the SCM activities required by subcontractors. The following information is provided as a supporting example.

> *[Project Abbreviation] subcontractor support will be identified by the [Project Abbreviation] Program Manager. Subcontractor support will be required to follow the process defined in this document. If a wavier is requested, the subcontractor must provide evidence of comparable configuration management procedures. These procedures will follow the same audit and control procedures described in Section xxx.*

Appendix A. Software Change Request Procedures

Refer to Appendix C, *Software Change Request Procedures,* of this text for example software Change Request Procedures.

Appendix B. [PROGRAM ABBREVIATION]
Software Organization

B.1 [PROGRAM ABBREVIATION] Organization

[PROGRAM ABBREVIATION] software elements can be categorized into three groups: module products, data products, and end-user products.

Module products and data products are used by software developers and not the users and are components of [PROGRAM ABBREVIATION].

Module products. Module products are software libraries that perform a specific set of functions. Module products provide reusable code ensuring consistency in function performance, eliminating duplication of effort in developing like functions, and reducing the amount of code. Module products are used in our end-user products and may be available for use by other companies/agencies.

Data products. Data products are database data files. Data products allow a single development, maintenance, and testing source, and provides consistent data and format. Data products are included as part of our end-user products and may be available for use by other companies/agencies. Data products may be sent to users as a product itself.

End-user products. End-user products are executable programs for users.

Configuration management of module products is accomplished according to this document. Each module product has an individual baseline and version numbers. Module product managers specify the version of the module product to be used in end-user products.

B.1.1 [PROGRAM ABBREVIATION] Module Products

A PVCS directory structure for all module products exists for each software product version released.

B.1.2 [PROGRAM ABBREVIATION] Data Products

B.1.3 [PROGRAM ABBREVIATION] Deliverables

B.1.4 [PROGRAM ABBREVIATION] Support Software

QUALITY ASSURANCE

The quality assurance (QA) process provides adequate assurance that the software products and processes in the project life cycle conform to their specified requirements and adhere to their established plans. The QA process must assure that each process, activity, and task required by the contract or described in plans are being performed in accordance with the contract and with those plans. QA also provides assurances that each software product has undergone software product evaluation, testing, and problem resolution. The quality assurance process may make use of the results of other supporting processes, such as verification, validation, joint review, audit, and problem resolution. Table 6-5 provides a list of the quality assurance process objectives.

ISO 9001 Goals

ISO 9001 describes requirements in support of the IEEE 12207 quality assurance process in three clauses for planning at both organizational and project levels. Clause 5.4.2, Quality Management System Planning, requires that upper management ensure that planning for the quality management system meet general requirements, as well as quality objectives, and that it maintain system integrity when changes are planned and implemented. This planning defines the content of software quality assurance plans.

Clause 7.3.1, Design and Development Planning, requires the planning and control of software product design and development. The interfaces between the different involved groups must be managed to ensure effective communication and the clear assignment of responsibility, including quality assurance. The quality assurance planning output must be updated, as appropriate, during design and development.

Clause 8.2.2, Internal Audit, requires that periodic internal audits must be conducted to determine if the quality management system conforms to planned arrangements and is effectively implemented and maintained. The software quality assurance plan (SQAP) includes the internal audit program that uses impartial and objective auditors for permit objective evaluations.

Software Quality Assurance Plan

The information provided here is designed to facilitate the definition of processes and procedures relating to software quality assurance activities. This guidance was developed using IEEE Std 12207.0, Guide to Lifecycle Processes [37], and IEEE

Table 6-5. Quality assurance process objectives

a) Identify, plan, and schedule quality assurance activities for the process or product.

b) Identify quality standards, methodologies, procedures, and tools for performing quality assurance activities and tailor to the project.

c) Identify resources and responsibilities for the performance of quality assurance activities.

d) Establish and guarantee the independence of those responsible for performing quality assurance activities.

e) Perform the identified quality assurance activities in line with the relevant plans, procedures, and schedules.

f) Apply organizational quality management systems to the project.

Std 730, IEEE Standard for Software Quality Assurance Plans, which has been adapted to support ISO 9001 requirements. The modification of the recommended SQAP table of contents to support the goals of ISO 9001 more directly is shown in Table 6-6. Additional information is provided in Appendix C, Software Process Work Products, which presents a recommended minimum set of software reviews and an example SQA inspection log.

Software Quality Assurance Plan Document Guidance

The following provides section-by-section guidance in support of the creation of a SQAP. The SQAP should be considered to be a living document. The SQAP should change to reflect any process improvement activity. This guidance should be used to help define a software quality process and should reflect the actual processes and

Table 6-6. Software quality assurance plan document outline

Title Page
Revision Page
Table of Contents
1. Introduction
    1.1 Purpose
    1.2 Scope
    1.3 Definitions, Acronyms, and Abbreviations
    1.4 References
2. SQA Management
    2.1 Organization
    2.2 Tasks
    2.3 Responsibilities
3. SQA Documentation
    3.1 Development, Verification and Validation, Use, and Maintenance
    3.2 Control
4. Standards and Practices
    4.1 Coding/Design Language Standards
    4.2 Documentation Standards
5. Reviews and Audits
6. Configuration Management
7. Testing
    7.1 System
    7.2 Integration
    7.3 Unit
8. Software Measures
9. Problem Tracking
10. Records Collection, Maintenance, and Retention
11. Training
12. Risk Management

procedures of the implementing organization. All information is provided for illustrative purposes only. Additional information is provided in the document template, *Software Quality Assurance Plan.doc,* which is located on the companion CD-ROM.

Introduction

This section should provide the reader with a basic description of the project to include associated contract information and any standards used in the creation of the document. The following is provided as an example:

> *The [Project Name] ([Proj Abbrev]) software products are developed under contract # [contract number], subcontract # [sub number], controlled by [customer office designator] at [customer location]. This Software Quality Assurance Plan (SQAP) has been developed to ensure that [project abbreviation] software products conform to established technical and contractual requirements. These may include, but are not limited to [company name] standards for Software Quality Assurance Plans.*

As described by IEEE Std 610.12, software quality assurance (SQA) is "a planned and systematic pattern of all actions necessary to provide adequate confidence that an item or product conforms to established technical requirements" (IEEE Std. 610.12, Standard Glossary of Software Engineering). This definition could be interpreted somewhat narrowly. The ISO 9001 supports the notion that quality assurance is performed on both processes and products and the 610.12 definition mentions only an item or product. This definition should be expanded to read "a planned and systematic pattern of all actions necessary to provide adequate confidence that an item, product, or process conforms to established technical requirements."

Purpose

This section should describe the purpose of the SQAP. It should list the name(s) of the software items covered by the SQAP and the intended use of the software. An example is provided.

> *The purpose of this SQAP is to establish, document, and maintain a consistent approach for controlling the quality of all software products developed and maintained by the [project abbreviation] Group. This plan defines the standardized set of techniques used to evaluate and report upon the process by which software products and documentation are developed and maintained.*
>
> *The SQA process described applies to all development, deliverables, and documentation maintained for the [project abbreviation] project. The SQA process and how it is implemented in the production, or maintenance life cycle, of each software product is shown in Figure xx.*

Scope

This section should describe the specific products and the portion of the software life cycle covered by the SQAP for each software item specified. The following is an example of scope.

The specific products developed and maintained by the [PROJECT AB-BREVIATION] development team may be found in the [project abbreviation] Software Configuration Management Plan.

In general, the software development products covered by this SQA Plan include, but are not limited to, the following:

- *source files*
- *libraries*
- *executable files*
- *data files*
- *makefiles*
- *link files*
- *documentation*

The software deliverables covered by this SQA Plan are:

- *[Project name] System Software*
- *Version Description Documents (VDDs)*
- *User's Manuals*
- *Programmer's Manuals*
- *Software Design Documents*
- *Software Requirements Specifications*

[PROJECT ABBREVIATION] software development tasks cover the full spectrum of software life cycle stages from requirements to system testing. As a result, Software Quality Assurance (SQA) activities must be adapted as needed for each specific software task.

Because of the cumulative and diverse nature of [PROJECT ABBRE-VIATION] software, this life cycle often includes the use of programs or subprograms that are already in use. These programs are included in the Preliminary Design, test documentation, and testing. They are included in the Detailed Design strictly in the design or modification of their interfaces with the remainder of the newly developed code.

In many of the [PROJECT ABBREVIATION] software products, changes are required to existing code, either as corrections or upgrades. The life cycle of software change is the same as the life cycle of new software, although certain steps may be significantly shorter. For example, system testing may omit cases that do not exercise the altered code.

The decision as to whether an individual task involves a new software development or a change to an existing program will be the responsibility of the Project Lead, Program Manager, Customer, associated CCB, or a combination thereof.

The life cycle for a specific project would be described in the associated software project management plan (SPMP). An example of a project life cycle and information in support of SQA activities is provided in Appendix C, Example Life Cycle, of this text.

SQA Management

This section should describe the project's quality assurance organization, its tasks, and its roles and responsibilities that will be used to effectively implement process and product quality assurance. A reference to any associated software project management plan should be provided here. The following is an example paragraph:

> *The [PROJECT ABBREVIATION] Program Manager, each [PROJECT ABBREVIATION] Project Team Lead, and the [PROJECT ABBREVIATION] Software Quality Assurance Manager (SQAM) will work together with the [Company Name] Software Engineering Process Group to develop, maintain, and implement effective software quality assurance management.*

Organization

This section should depict the organizational structure that influences and controls the quality of the software. This should include a description of each major element of the organization together with the roles and delegated responsibilities. The amount of organizational freedom and objectivity to evaluate and monitor the quality of the software, and to verify problem resolutions, should be clearly described and documented. In addition, the organization responsible for preparing and maintaining the SQAP should be identified. A typical organizational breakdown follows.

> *Program Manager. The [PROJECT ABBREVIATION] Program Manager (PM) is ultimately responsible for ensuring the SQA process is developed, implemented, and maintained. In this capacity, the PM will work with the [COMPANY NAME] SQAM and the [PROJECT ABBREVIATION] SQAM in establishing standards and implementing the corporate SQAP as it applies to the [PROJECT ABBREVIATION] program.*
>
> *The [PROJECT ABBREVIATION] PM is ultimately responsible for ensuring that the Project Lead(s) and their associated project team(s) are trained in SQA policies and procedures that apply to their function on each team. This will consist of initial training for new members and refresher training, as required, due to process or personnel changes.*
>
> *Project Lead. The [PROJECT ABBREVIATION] Project Lead (PL) shall provide direct oversight of SQA activities associated with project software development, modification, or maintenance tasks assigned within project functional areas.*
>
> *The Project Lead(s) are responsible for ensuring that project software quality assurance processes are followed. Similarly, the PL will ensure*

that all members of their project team regularly implement the SQA poli-cies and procedures that apply to their specific function on the team.

The PL is responsible for implementing the actions of this SQA process for all [PROJECT ABBREVIATION] software products, includ-ing end-user products, module products, and data. The PL will be respon-sible for controlling the product, ensuring that only approved changes are implemented. Each PL will maintain this process, with all changes to the process approved by the PM and associated personnel.

Tasks

This section should describe the portion of the software life cycle covered by the SQAP. All tasks to be performed, the entry and exit criteria for each task, and rela-tionships between these tasks and the planned major checkpoints should also be de-scribed. The sequence and relationships of tasks, and their relationship to the pro-ject management plan master schedule, should also be indicated. The following provides a content example.

The [PROJECT ABBREVIATION] PM shall ensure that a planned and systematic process exists for all actions necessary to provide software products that conform to established technical and contractual require-ments.

The following SQA checkpoints will be placed in the software develop-ment life cycle:

   a. *At the end of the Software Requirements and Software System Test Planning activities, there shall be a Requirements Review in accor-dance with Section xxx of this document.*
   b. *At the end of the Preliminary Design activity, there shall be a Pre-liminary Design Review in accordance with Section xxx of this doc-ument.*
   c. *At the end of Detailed Design, there shall be a Critical Design Re-view in accordance with Section xxx of this document.*
   d. *At the end of Unit Testing, there shall be a walk-through in accor-dance with Section xxx of this document.*
   e. *At the end of Integration Testing, there shall be a Test Review in accordance with Section xxx of this document.*
   f. *At the end of System Testing, there shall be a Test Review and an audit in accordance with Section xxx of this document.*
   g. *At Release, there shall be an SQA review using the checklists found in Appendix xx.*

The following SQA checkpoints will be placed in the software mainte-nance life cycle:

   a. *At the end of the Analysis activity there shall be a Requirements Review in accordance with Section xxx of this document.*

    b. At the end of Design, there shall be a Preliminary Design Review in accordance with Section xxx of this document.

    c. At the end of Implementation, there will be an walk-through in accordance with procedures identified in the associated System Test Plan and in Section xxx of this document.

    d. At Release, there shall be an SQA review using the release checklists found in Appendix xx.

    e. Acceptance Testing is dependant upon criteria determined by the customer. This criterion is based upon the documentation released with the software (i.e., the SRS, VDD, user's documentation, etc.).

Each PL, or a designee, shall evaluate all software products to be delivered to the customer during the course of any project. This evaluation must ensure that the products are properly annotated, are adequately tested, are of acceptable quality, and are updated to reflect changes specified by the customer. Additionally, the PL is responsible for ensuring that both the content and format of the software documentation meet customer requirements.

Responsibilities

This section should identify the specific organizational element that is responsible for performing each task as shown by the following.

Project Lead Responsibilities. The [PROJECT ABBREVIATION] PM has overall responsibility for [PROJECT ABBREVIATION] program SQA responsibilities.

The [PROJECT ABBREVIATION] SQAM has the overall responsibility for monitoring compliance with the project procedures, noting any deviations and findings reporting.

The [PROJECT ABBREVIATION] PLs have the responsibility for ensuring that individual team members are trained in Project SQA procedures and practices. They may delegate the actual training activities to [PROJECT ABBREVIATION] SQAM. Project Leads are also responsible or reporting changes in Project procedures to the [PROJECT ABBREVIATION] SQAM.

Each individual [PROJECT ABBREVIATION] employee shall be responsible for learning and following the SQA procedures and for suggesting changes to their PL when necessary. Each PL shall be responsible for the implementation and accuracy of all SQA documentation and life cycle tracking.

Configuration Control Board (CCB). Each baseline software product developed by [PROJECT ABBREVIATION] has an associated Configuration Control Board (CCB). The associated CCB has the authority to approve, disapprove, or revise all Software Quality Assurance reporting procedures.

An example of the delineation of SQA life cycle responsibility is provided in Appendix C, Example Life Cycle, of this text.

SQA Documentation

This section should identify the documentation governing the development, verification and validation, use, and maintenance of the software. List all documents that are to be reviewed or audited for adequacy identifying the reviews or audits to be conducted. Refer to IEEE Std 730-2002 for detailed information supporting SQA documentation requirements.

Development, Verification and Validation, Use, and Maintenance

This section should address how project documentation is developed, verified, validated, and maintained. The following is an example excerpt in support of this requirement:

Control

This section should describe the responsibilities associated with software project documentation. The following provides an example of the type of information to support this requirement.

> *The [PROJECT ABBREVIATION] ([PROJECT ABBREVIATION]) Program Manager (PM), Project Lead(s), and team members shall share responsibility for SQA documentation. Document reviews and integration of any required documentation changes will occur as specified in Section 2.3.1 of this document. Refer to [PROJECT ABBREVIATION] Software Configuration Management Process for additional information regarding specific product (i.e., software, documentation . . .) maintenance.*
>
> *Changes to requirements, software baselines and completed maintenance items will be reported in the deliverable's Version Description Document. Status reports will be provided to [Customer Name] no less than bi-weekly.*

Standards and Practices

This section should identify the standards, practices, conventions, quality requirements, and measurements to be applied. Product and process measures should be included in the measurements used and may be identified in a separate measurements plan. This section should also describe how conformance with these items will be monitored and assured. The following provides an example of typical information.

> *The following standards shall be applied to all programs and incorporated into existing programs whenever and wherever possible. Retrofitting of existing programs to meet new standards will be avoided when this retrofitting impacts production.*
>
> *Coding/Design Language Standards. All software developed and maintained by [PROJECT ABBREVIATION] project teams shall adhere to [PROJECT ABBREVIATION] Programming Standards , [Additional Coding Reference]. Deviations from the Standards shall be noted on the SCR form or recorded as a CER.*
>
> *Documentation Standards. All [PROJECT ABBREVIATION] documentation shall be adapted from the [Additional Documentation Reference] and IEEE documentation guidelines.*

Reviews and Audits

This section should describe all software reviews to be conducted. These might include managerial reviews, acquirer–supplier reviews, technical reviews, inspections, walk-throughs, and audits. The schedule for software reviews as they relate to the software project's schedule should also be listed, or reference made to a software project management plan containing the information. This section should describe how the software reviews would be accomplished.

Configuration Management

This section shall identify associated configuration management activities necessary to meet the needs of the SQAP. An example follows.

> *For information regarding configuration control and status accounting procedures refer to the [PROJECT ABBREVIATION] Configuration Management Plan. Configuration Management Maintenance is the responsibility of each [PROJECT ABBREVIATION] PL and each project team Configuration Manager.*
>
> *The Project SQA Manager shall audit the baseline versions of the software product and accompanying documentation prior to software release for validation. (Please refer to Appendix xx, Software Checklists). The [PROJECT ABBREVIATION] PM will verify that a baseline is established and preserved and that documented configuration management policies and [COMPANY NAME] corporate policies are implemented.*

Testing

This section should identify all the tests not included in the software verification and validation plan for the software covered by the SQAP and should state the methods to be used. If a separate test plan exists, it shall be referenced. An example follows:

> *Testing of software developed, modified and/or maintained under the [PROJECT ABBREVIATION] contract must be tailored to meet the requirements of the specific project Work Request. Typical testing efforts include unit, integration, and system testing. The [PROJECT ABBREVIATION] Program Manager will ensure that all project software testing is planned and executed properly. Please refer to the [PROJECT ABBREVIATION] Configuration Management Process and program test plans, listed in Section xxx.*

The system test program, also referred to as DT&E testing, should be broad enough to ensure that all the software requirements are met. The software test plan is to be written at the same time that the software requirements are being finalized. It may be useful to draft a test plan while the requirements are being defined. As each requirement is defined, a way to determine that it is satisfied should be delineated. This method is then translated into a test case for the system testing and is added to the software test plan.

System Testing. The system testing should consist of a list of test cases (descriptions of the data to be entered) and a description of the resulting program output.

Integration Testing. The integration testing should check to ensure that the modules work together as specified in the SDD. Integration testing should be included as part of the software test plan to ensure that all the modules to be tested are executed and exercised thoroughly.

Unit Testing. Unit testing is the detailed test of a single module to ensure that it works as it should. Unit testing should be identified in the software test plan to ensure that every critical function in the module is executed at least once. Critical functionality should be identified in the software test plan.

Software Measures

This section should describe the collection, maintenance, and retention of measures associated with the software quality assurance process. A separate software measurement and measures plan is often used, and if so, a reference to this plan should be provided. The following is an example of the information supporting this section of the document:

> *The [PROJECT ABBREVIATION] ([PROJECT ABBREVIATION]) SQA Manager shall ensure that [PROJECT ABBREVIATION] Software projects follow the guidelines established by the [PROJECT ABBREVIATION] Software Measurement and Measures Plan.*

The SQA Manager shall ensure the project measures are appropriate and will provide recommendations for improvements on the project's measurement process, as required.

The following software measures will be recorded and used in support of the [PROJECT ABBREVIATION] SQA effort:

a) *A history of the bugs and enhancements, a list of functions which are changed to fix the bugs or implement the enhancements, and the time and effort required to fix the bugs or implement the enhancements. The [PROJECT ABBREVIATION] Software Change Request Tracking System will provide such information (CER/SCR System).*

b) *Total numbers of errors, opened versus closed; errors opened and closed since last report. Errors being defined as:*
   1) *Requirements errors—Baseline Change Request (BCR)*
   2) *Coding defects—Test Problem Reports (TPRs)*
   3) *Documentation defects—Interface Change Notice (DCR)*
   4) *Bad Fixes—Internal Test Report (ITR)*

c) *Total implemented requirements versus planned requirements.*

Problem Tracking

This section should describe the practices and procedures to be followed for reporting, tracking, and resolving problems or issues identifieded in both software items and the software development and maintenance process. It should also describe organizational responsibilities concerned with their implementation. The following provides some example text.

Please refer to [PROJECT ABBREVIATION] Configuration Management Process. This document describes the practices and procedures followed for reporting, tracking, and resolving problems identified in software items and the associated development, maintenance, and organizational responsibilities.

Deficiencies in the SQA Plan shall be corrected using the following procedure:

1) *A deficiency is noted, either by a [COMPANY NAME] employee finding the procedure difficult to follow or by the SQA manager noting that the procedure has not been followed.*

2) *The PM, or a designee, notes the deficiency and develops a way to correct it. In this, he shall have help from the employee who had trouble following the procedure or who did not follow the procedure.*

3) *A new version of the SQA Plan is written, printed, and disseminated to [PROJECT ABBREVIATION] Program personnel and [COMPANY NAME] Corporate SQA. The new version shall incorporate the correction.*

Records Collection, Maintenance, and Retention

This section should identify all SQA documentation and describe how the documentation is maintained, while designating the retention period. The following is an example.

> *The Software Developmental Library (SDL) for [PROJECT ABBREVIA-TION] program software may be found in Appendix A. Documents pertaining to developmental configuration, which are also maintained, may be found in Section 3.1. Access to all documents is controlled by each project team's Project Lead, or designee.*
>
> *This documentation will be maintained in [Maintenance location and address]. All documentation will be maintained on in a shared [PROJECT ABBREVIATION] subdirectory [PROJECT ABBREVIATION]/ [path]/shared/<document name>. Weekly backups of this system occur and are stored off-site.*
>
> *All documentation will be maintained throughout the length of the [PROJECT ABBREVIATION] contract.*

Training

This section should identify the training activities necessary to meet the needs of the SQAP. The following is an example:

> *It is the responsibility of the [PROJECT ABBREVIATION] Program Manager (PM) to allow for the training necessary to accomplish SQA goals and implementation. Similarly, the PL will ensure that all members of the project team are trained in the SQA policies and procedures that apply to their function on each team. This will consist of initial training for new members and refresher training, as required, due to process or personnel changes.*

Risk Management

This section should specify all methods and procedures employed to identify, assess, monitor, and control areas of risk arising during the portion of the software life cycle covered by the SQAP. Reference to the software project management plan, or a separate risk management plan, would suffice here, as shown by the following text.

> *Program risks will be identified either by individual team members or CCB members. The risk tracking form (refer to [PROJECT ABBREVIA-TION] Software Development Plan) will be used to document and track all risks identified. All risks will be submitted to the [PROJECT ABBRE-VIATION] CCB at weekly staff meetings to determine the severity and probability. Risks determined to be significant will be assigned an owner. The owner will determine the plan for dealing with the risk and propose a*

plan to the [PROJECT ABBREVIATION] CCB. The owner will track the risk to closure. Risks that are not deemed significant will be canceled.

The guidelines used by the [PROJECT ABBREVIATION] CCB to define risk severity and probability is listed in the [PROJECT ABBREVIATION] Software Development Plan.

Additional information is provided in the *Software Quality Assurance Plan.doc* document template on the companion CD-ROM. Appendix C, Software Process Work Products, of this book also provides several examples of supporting process artifacts.

VERIFICATION

The verification process determines whether the software products of an activity fulfill the requirements or conditions imposed on them in the previous activities. This process may include analysis, review, and test. Verification should be integrated, as early as possible, with the process that employs it. Verification is defined as confirmation by examination and provision of objective evidence that specified requirements have been fulfilled. Table 6-7 provides a list of the verification process objectives.

ISO 9001 Goals

ISO 9001 describes requirements in support of the IEEE 12207 verification process. The intent of ISO 9001 is to use robust processes to design quality into the product and thus reduce the verification costs of testing quality into the product. The key requirement is Clause 7.3.5, Design and Development Verification, to ensure the design and development output meets the design and development input requirements, when performed in accordance with planned arrangements in Clause 7.3.1, Design and Development Planning, which concerns planning and controlling the verification activities for each activity output in the product design and development cycle. Verification may comprise reviews of output (e.g., by inspections and walk-throughs); analysis; demonstrations, including prototypes; simulations; or tests. The verification results and subsequent follow-up actions must be recorded. Complementing Clause 7.3.5 is Clause 7.3.3, Design and Development Outputs, which requires the design and development process outputs be documented to enable verification against the inputs to the process. Complementing Clause 7.3.1 is Clause

Table 6-7. Verification process objectives

a) Identify criteria for verification of all required work products.
b) Perform requirements verification activities.
c) Find and remove defects from products produced by the project.

7.3.7, Control of Design and Development Changes, which requires design and development changes to be reviewed, verified, validated (as appropriate), and approved before implementation.

Within the larger scope of product realization, Clause 7.1, Planning of Product Realization, processes must be planned and developed. The planning must determine the verification, validation, monitoring, inspection, and test activities. This verification plan may cover customer requirements; purchased products; measuring, monitoring, or testing devices; internal audit actions implementation, product; and nonconforming product. In summary, the six ISO 9001 clauses supporting Verification are:

Clause 7.2.2, Review of Requirements Related to the Product, requires software product requirements review before committing to supply the software product to the customer. The results of the review, and any subsequent follow-up actions, must be maintained.

Clause 7.4.3, Verification of Purchased Product, is required to ensure that the purchased products meet the specified purchase requirements. The purchased product complexity can range from COTS to subcontracted software development or to system hardware. Thus, the organization needs to determine the methods by which verification, validation, and acceptance of the purchased product will be achieved.

Clause 7.6, Control of Measuring and Monitoring Devices, requires validation of monitoring and measurements devices or testing devices.

Clause 8.2.2, Internal Audit, concludes with audit findings that require follow-up actions, which must be verified.

Clause 8.2.4, Monitoring and Measurement of Product, requires monitoring and measuring product characteristics to verify that product requirements are being met during appropriate product realization stages according to the plans.

Clause 8.3, Control of Nonconforming Product, ensures that any nonconforming software product or configuration item (CI) is identified and controlled to prevent its unintended use or delivery. When the nonconformity is corrected, it must be reverified to show conformity.

Inspections

The information provided here is based upon the recommendations provided in IEEE Std 1028, IEEE Standard for Software Reviews. Inspections are an effective way to detect and identify problems early in the software development process. Inspections should be peer-driven and should follow a defined, predetermine practice. The criteria for inspections are presented here to help organizations define their inspection practices. IEEE Std 1044, IEEE Standard Classification for Software Anomalies [13], provides valuable information regarding the categorization of software anomalies. Additional information is provided in Appendix C, which presents a recommended minimum set of software reviews and an example SQA inspection log.

Introduction

Software inspections can be used to verify that a specified software product meets its requirements. These requirements can address functional characteristics, quality attributes, or adherence to regulations or standards. Any inspection practice should collect anomaly data used to identify trends and improve the development process. See Classic Anomaly Class Categories in Appendix C, Work Products.

Examples of software products subject to inspections are:

a) Software requirements specification
b) Software design description
c) Source code
d) Software test plan documentation
e) Software user documentation
f) Maintenance manual
g) System build procedures
h) Installation procedures
i) Release notes

Responsibilities

Each inspection should have a predefined set of roles and responsibilities; the assignment of these responsibilities should occur on a case-by-case basis and is dependant upon the item subject to inspection.

Participants

Inspections should consist of three to six participants and should be led by an impartial facilitator who is trained in inspection techniques. Determination of remedial or investigative action for an anomaly is a mandatory element of a software inspection, although the resolution should not occur in the inspection meeting. Collection of data for the purpose of analysis and improvement of software engineering procedures (including all review procedures) is strongly recommended but is not a mandatory element of software inspections. Inspection participants may include:

Inspection Leader. The inspection leader is responsible for administrative tasks pertaining to the inspection and is responsible for the planning and preparation of the inspection. The leader is responsible for ensuring that the inspection is conducted in an orderly manner and meets its objectives. The leader is responsible for the collection of inspection data and the issuance of the inspection report.

Recorder. The recorder is responsible for recording all inspection data required for process analysis. The recorder should document all anomalies, action items, decisions, and recommendations made by the inspection team. The inspection leader may also act as recorder.

Reader. The reader should methodically lead the inspection team through the software product, providing a summary, when appropriate, highlighting all items of interest.

Author. The author contributes the product for inspection and participates when a special understanding of the software product is required. The author is responsible for any rework required to make the software product meet inspection exit criteria.

Inspector. All participants in the review are inspectors. The author should not act as inspection leader and should not act as reader or recorder. All other roles may be shared among the team members and participants may act in more than one role. Individuals holding management positions over any member of the inspection team should not participate in the inspection. Inspectors are responsible for the identification and description of all anomalies found in the software product. Inspectors should represent different viewpoints at the meeting (for example, sponsor, requirements, design, code, safety, test, independent test, project management, quality management, and hardware engineering). Only those viewpoints with any relevance to the product should be present. Each inspector should be assigned a specific review topic to ensure effective product coverage. The inspection leader should assign these roles prior to the inspection.

Input

Input to the inspection should include the objectives, identification of the software product, the inspection procedure, all reporting forms, all guidelines or standards, and any known anomalies or issues.

Authorization

Inspections should be identified in any associated software project management plan or software verification and validation plan, and should be reflected in project scheduling and resource allocation.

Preconditions

An inspection should be conducted only when a statement of objectives for the inspection is established and the required inspection inputs are available.

Minimum Entry Criteria

An inspection should not be conducted until the software product is complete and conforms to project standards for content and format, all tools and documentation required in support of the inspection are available, and any related prior milestones have been satisfied. If the inspection is a reinspection, all previously identified problems should have been resolved.

Management Preparation

Managers are responsible for planning the time and resources required for inspection, including support functions, and assuring that these are identified in the associated software project management plan. Managers should also ensure that individuals receive adequate training and orientation on inspection procedures, and possess appropriate levels of expertise and knowledge sufficient to comprehend the software product under inspection.

Planning the Inspection

The author of the software product is responsible for the assembly of all inspection materials. The inspection leader is responsible for the identification of the inspection team, the assignment of specific responsibility, and scheduling and conducting the inspection. As a part of the planning procedure, the inspection team should determine if anomaly resolution is to occur during the inspection meeting or if resolution will occur under other circumstances.

Overview of Inspection Procedures

The author of the software product should present a summary overview for the inspectors. The inspection leader should answer questions about any checklists and assign roles. Expectation should be reviewed, such as the minimum preparation time anticipated and the typical number of anticipated anomalies.

Preparation

Each inspection team member should examine the software product and document all anomalies found. All items found should be sent to the inspection leader. The inspection leader should classify (i.e., consolidate) all anomalies and forward these to the author of the software product for disposition. The inspection leader, or reader, should specify the order of the inspection (e.g., hierarchical). The reader should be prepared to present the software product at the inspection meeting.

Examination

The inspection meeting should follow this agenda:

> *Introduce the meeting.* The inspection leader should introduce all participants and review their roles. The inspection leader should state the purpose of the inspection and should remind the inspectors to focus their efforts toward

anomaly detection, direct remarks to the reader and not the author, and to comment only on the software product and not the author. See Classic Anomaly Class Categories in Appendix C.

Establish preparedness. The inspection leader should verify that all participants are prepared. The inspection leader should gather individual preparation times and record the total in associated inspection documentation.

Review general items. General, nonspecific, or pervasive anomalies should be presented first. If problems are discovered that inhibit the further efficient detection of error, the leader may choose to reconvene the inspection following the resolution of all items found.

Review software product and record anomalies. The reader should present the software product to the inspection team. The inspection team should examine the software product and create an anomaly list of items found. The recorder should enter each anomaly, location, description, and classification on the anomaly list (IEEE Std 1044 may be used to support anomaly classification). See Inspection Log Defect Summary Description in Appendix C.

Review the anomaly list. At the end of the inspection meeting, the team should review the anomaly list to ensure its completeness and accuracy. The inspection leader should allow time to discuss every anomaly for which disagreement occurred.

Make exit decision. The purpose of the exit decision is to bring an unambiguous closure to the inspection meeting. Common exit criteria are listed below:

a) *Accept with no or minor rework.* The software product is accepted as is or with only minor rework (requires no further verification).

b) *Accept with rework verification.* The software product is to be accepted after the inspection leader or a designated member of the inspection team verifies rework.

c) *Reinspect.* Schedule a reinspection to verify rework. At a minimum, a reinspection should examine the software product areas changed to resolve anomalies identified in the last inspection, as well as the side effects of those changes.

Exit Criteria

An inspection is considered complete when prescribed exit criteria have been met.

Output

The output of the inspection provides documented evidence that identifies that the project is being inspected. It provides a list of all team members, the meeting duration, a description of the product inspected, all inspection inputs, objectives, anomaly list and summary, a rework estimate, and total inspection preparation time.

Data Collection Recommendations

Inspection data should contain the identification of the software product, the date and time of the inspection, the inspection leader, the preparation and inspection times, the volume of the materials inspected, and the disposition of the inspected software product. The management of inspection data should provide for the capability to store, enter, access, update, summarize, and report categorized anomalies.

Anomaly Classification

Anomalies may be classified by technical type. IEEE Std 1044 provides effective guidance in support of anomaly classification. Appendix A of IEEE Std 1044 provides sample screens from an anomaly reporting system. The data fields for this system are listed in Appendix A under the Verification Artifacts section. See Classic Anomaly Class Categories in Appendix C.

Anomaly Ranking

Anomalies may be ranked by potential impact on the software product. Three ranking categories are provided as examples:

Category 1. Major anomaly of the software product, or an observable departure from specification, with no available workaround.

Category 2. Minor anomaly of the software product, or departure from specification, with existing workaround.

Category 3. Cosmetic anomalies that deviate from relevant specifications but do not cause failure of the software product.

The information is captured in the Inspection Report Description and the Inspection Log Defect Summary Description, as shown in Appendix C.

Improvement

Inspection data should be analyzed regularly in order to improve the inspection and software development processes. Frequently occurring anomalies may be included in the inspection checklists or role assignments. The checklists themselves should also be evaluated for relevance and effectiveness.

Walk-throughs

The following information is based on IEEE Std 1028, IEEE Standard for Software Reviews; and IEEE Std 1044, IEEE Guide to Classification of Software Anomalies, which have been adapted to support ISO 9001 requirements.

The purpose of a systematic walk-through is to evaluate a software product. Walk-throughs effectively identify anomalies and contribute toward the improvement of the software development process. Walk-throughs are also effective training activities, encouraging technical exchanges and information sharing. Examples of software products subject to walk-throughs include software project documentation, source code, and system build or installation procedures.

Responsibilities

The following roles should be established:

> *Leader.* The walk-through leader is responsible for administrative tasks pertaining to the walk-through and is responsible for the planning and preparation of the walk-through. The leader is responsible for ensuring that the walk-through is conducted in an orderly manner and meets its objectives. The leader is responsible for the collection of walk-through data and the issuance of the summary report.
>
> *Recorder.* The recorder should methodically document all decisions, actions, and anomalies discussed during the walk-through.
>
> *Author.* The author contributes the product for walk-through and participates when a special understanding of the software product is required. The author is responsible for any rework required to address all anomalies identified.
>
> *Team Member.* All participants in the walk-through are team members. All other roles may be shared among the team members, and participants may act in more than one role. Individuals holding management positions over any member of the walkthrough team should not participate in the walk-through.

Input

Input criteria should include the walk-through objectives, the software product, anomaly categories (IEEE Std 1044 may be used to support anomaly classification), and all relevant standards. See Classic Anomaly Class Categories in Appendix C.

Authorization

Walk-throughs should be identified in any associated software project management plan or software verification and validation plan, and should be reflected in project scheduling and resource allocation.

Preconditions

A walk-through should be conducted only when a statement of objectives for the walk-through is established and the required walk-through inputs are available.

Management Preparation

Managers are responsible for planning the time and resources required for the walk-through, including support functions, and ensuring that these are identified in the associated software project management plan. Managers should also ensure that individuals receive adequate training and orientation on walk-through procedures and they possess appropriate levels of expertise and knowledge sufficient to evaluate the software product.

Planning the Walk-Through

The walk-through leader is responsible for identifying the walk-through team, distributing walk-through materials, and scheduling and conducting the walk-through.

Overview

The author of the software product should present a summary overview for the team.

Preparation

Each walk-through team member should examine the software product and prepare a list of items for discussion. These items should be categorized as either specific or general, where general would apply to the entire product and specific to one part. All anomalies detected by team members should be sent to the walk-through leader for classification. The leader should forward these to the author of the software product for disposition. The walk-through leader should specify the order of the walk-through (e.g., hierarchical).

Examination

The walk-through leader should introduce all participants, describe their roles, and state the purpose of the walk-through. All team members should be reminded to focus their efforts toward anomaly detection. The walk-through leader should remind the team members to comment only on the software product and not its author. Team members may pose general questions to the author regarding the software product.

The author should then present an overview of the software product under review. This is followed by a general discussion during which team members raise their general items. After the general discussion, the author serially presents the software product in detail. Team members raise their specific items when the author reaches them in the presentation. It is important to note that new items may be raised during the meeting.

The walk-through leader is responsible for guiding the meeting to a decision or action on each item. The recorder is responsible for noting all recommendations and required actions. Appendix C, Work Products, contains several walk-through forms: Requirements Walk-through Form, Software Project Plan Walk-through Checklist, Preliminary Design Walk-through Checklist, Detailed Design Walk-through Checklist, Program Code Walk-through Checklist, Test Plan Walk-through Checklist, Walk-through Summary Report, and Classic Anomaly Class Categories.

After the walk-through meeting, the walk-through leader should issue a walk-through report detailing anomalies, decisions, actions, and other information of interest. This report should include a list of all team members, description of the software product, statement of walk-through objectives, list of anomalies and associated actions, all due dates and individual responsible, and the identification of follow-up activities.

Data Collection Recommendations

Walk-through data should contain the identification of the software product, the date and time of the walk-through, the walk-through leader, the preparation and walk-through times, the volume of the materials inspected, and the disposition of the software product. The management of walk-through data should provide for the capability to store, enter, access, update, summarize, and report categorized anomalies.

Anomaly Classification

Anomalies may be classified by technical type. IEEE Std 1044 provides effective guidance in support of anomaly classification. Appendix A of IEEE Std 1044 provides sample screens from an anomaly reporting system. The data fields for this system are listed in Appendix A in the Verification Artifacts section.

Anomaly Ranking

Anomalies may be ranked by potential impact on the software product. Three ranking categories are provided as examples:

Category 1. Major anomaly of the software product, or an observable departure from specification, with no available workaround.

Category 2. Minor anomaly of the software product, or departure from specification, with existing workaround.

Category 3. Cosmetic anomalies that deviate from relevant specifications but do not cause failure of the software product.

Improvement

Walk-through data should be analyzed regularly in order to improve the walk-through and software development processes. Frequently occurring anomalies may be included in the walk-through checklists or role assignments. The checklists themselves should also be evaluated for relevance and effectiveness.

VALIDATION

The validation process is a process for determining whether the requirements and the final system or software product fulfills its specific intended use. Validation is defined as confirmation by examination and provision of objective evidence that the particular requirements for a specific intended use are fulfilled. Validation is often performed by testing, conducted at several levels or approaches. Table 6-8 provides a list of the validation process objectives.

ISO 9001 Goals

ISO 9001 describes requirements in support of the IEEE 12207 validation process. The key requirement is Clause 7.3.6, Design and Development Validation, to confirm that the resulting product is capable of meeting the requirements for its specified application or intended use, where known and performed in accordance with planned arrangements in Clause 7.3.1, Design and Development Planning, which is the planning and controlling of the validation activities appropriate to each design and development stage. Complementing Clause 7.3.1 is Clause 7.3.7, Control of Design and Development Changes, which requires design and development changes to be reviewed, verified, validated (as appropriate), and approved before implementation.

Within the larger scope of product realization, Clause 7.1, Planning of Product Realization, processes must be planned and developed. The planning must determine the verification, validation, monitoring, inspection, and test activities. This validation plan may cover software replication and software installation in Clause 7.5.1, Control of Production and Service Provision, to plan and carry out production and service provision under controlled conditions. Note that Clause 7.5.2, Validation of Processes for Production and Service Provision, rarely applies to software engineering organizations.

Software Test Plan

The documentation supporting software testing should define the scope, approach, resources, and schedule of the testing activities. It should identify all items being tested, the features to be tested, the testing tasks to be performed, the personnel responsible for each task, and the risks associated with testing. IEEE Std 829, IEEE Standard for Software Test Documentation [5], and IEEE 12207.1, Standard for In-

Table 6-8. Validation process objectives

a) Identify criteria for validation of all required work products.
b) Perform required validation activities.
c) Provide evidence that the work products, as developed, are suitable for their intended use.

formation Technology—Software Life Cycle Processes—Life Cycle Data, were used as primary references for the information provided here.

The IEEE also publishes IEEE Std 1008, IEEE Standard for Software Unit Testing [8], and IEEE Std 1012, IEEE Standard for Software Verification and Validation [9], as part of their software engineering standards collection. These documents provide additional information in support of the development and definition of software test processes and procedures. Table 6-9 provides an example document outline. Additional information in support of the work products commonly associated with this ISO 9001 requirement may be found in Appendix C.

Software Test Plan Document Guidance

The following provides section-by-section guidance in support of the creation of a software test plan. It does not identify specific testing methodologies, approaches, techniques, facilities, or tools. As with any defined software engineering process, additional supporting documentation may be required. The development of a software test plan (STP) is integral to the software development process. Additional information is provided in the document template, *Software Test Plan.doc,* which is located on the companion CD-ROM.

Identification

This subsection should provide information that uniquely identifies the software effort and this associated test plan. This can also be provided in the form of a unique test plan identifier. The following text provides and example.

> *This software test plan is to detail the testing planned for the [Project Name] ([Project Abbreviation]) Version [xx], Statement of Work [date], Task Order [to number], Contract No. [Contract #], and Amendments.*
>
> *The goal of [project acronym] development is to [goal]. This software will allow [purpose].*
>
> *Specifics regarding the implementation of these modules are identified in the [Project Abbreviation] Software Requirements Specification (SRS) with line item descriptions in the accompanying Requirements Traceability Matrix (RTM).*

Document Overview

This subsection should provide a summary of all software items and software features to be tested. The need for each item and its history may be addressed here as well. References to associated project documents should be cited here. The following provides an example.

> *This document describes the Software Test Plan (STP) for the [PROJECT ABBREVIATION] software. [PROJECT ABBREVIATION] documenta-*

Table 6-9. Software test plan document outline

Title Page
Revision Page
Table of Contents
1. Introduction
2. Scope
    2.1 Identification
    2.2 System Overview
    2.3 Document Overview
    2.4 Acronyms and Definitions
        2.4.1 Acronyms
        2.4.2 Definitions
3. Referenced Documents
4. Software Test Environment
    4.1 Development Test and Evaluation
        4.1.1 Software Items
        4.1.2 Hardware and Firmware Items
        4.1.3 Other Materials
        4.1.4 Proprietary Nature, Acquirer's Rights, and Licensing
        4.1.5 Installation, Testing, and Control
        4.1.6 Participating Organizations
    4.2 Test Sites
        4.2.1 Software Items
        4.2.2 Hardware and Firmware Items
        4.2.3 Other Materials
        4.2.4 Proprietary Nature, Acquirer's Rights, and Licensing
        4.2.5 Installation, Testing, and Control
        4.2.6 Participating Organizations
5. Test Identification
    5.1 General Information
        5.1.1 Test Levels
        5.1.2 Test Classes
            5.1.2.1 Check for Correct Handling of Erroneous Inputs
            5.1.2.2 Check for Maximum Capacity
            5.1.2.3 User Interaction Behavior Consistency
            5.1.2.4 Retrieving Data
            5.1.2.5 Saving Data
            5.1.2.6 Display Screen and Printing Format Consistency
            5.1.2.7 Check Interactions between Modules
            5.1.2.8 Measure Time of Reaction to User Input
            5.1.2.9 Functional Flow
        5.1.3 General Test Conditions
        5.1.4 Test Progression
    5.2 Planned Tests
        5.2.1 Qualification Test
        5.2.2 Integration Tests
        5.2.3 Module Tests
        5.2.4 Installation Beta Tests   *(continued)*

Table 6-9. *Continued*

6. Test Schedules
7. Risk Management
8. Requirements Traceability
9. Notes
APPENDIX A. Software Test Requirements Matrix
APPENDIX B. Qualification Software Test Description
APPENDIX C. Integration Software Test Description
APPENDIX D. Module Software Test Description

tion will include this STP, the [PROJECT ABBREVIATION] Software Requirements Specification (SRS), the Requirements Traceability Matrix (RTM), the [PROJECT ABBREVIATION] Software Development Plan (SDP), the Software Design Document (SDD), the [PROJECT ABBREVIATION] User's Manual, the [PROJECT ABBREVIATION] System Administrator's Manual, and the [PROJECT ABBREVIATION] Data Dictionary.

This STP describes the process to be used and the deliverables to be generated in the testing of the [PROJECT ABBREVIATION]. This plan, and the items defined herein, will comply with the procedures defined in the [PROJECT ABBREVIATION] Software Configuration Management (SCM) Plan and the [PROJECT ABBREVIATION] Software Quality Assurance (SQA) Plan. Any nonconformance to these plans will be documented as such in this STP.

This document is based on the [PROJECT ABBREVIATION] Software Test Plan template with tailoring appropriate to the processes associated with the creation of the [PROJECT ABBREVIATION]. The information contained in this STP has been created for [Customer Name] and is to be considered "For Official Use Only."

Please refer to section "6. Schedule" in the SDP for information regarding documentation releases.

Acronyms and Definitions

All relevant acronyms and definitions should be included in this subsection.

Referenced Documents

This section should include all material referenced during the creation of the test plan.

Development Test and Evaluation

Provide a high-level summary of all key players, facilities required, and site of test performance. The following is provided as an example:

> *Qualification, integration, and module level tests are to be performed at [company name], [site location]. All testing will be conducted in the development center [room number]. The following individuals must be in attendance: [provide list of performers and observers].*

Software Items

This subsection should provide a complete list of all software items used in support of testing, including versions. If the software is kept in an online repository, reference to the storage location may be cited instead of listing all items here. The following is an example.

> *Software used in the testing of [PROJECT ABBREVIATION]:*
> *   [Software List]*
> *   For details of client and server software specifications see the [PROJECT ABBREVIATION] Software Requirements Specifications (SRS).*

Hardware and Firmware Items

This subsection should provide a complete list of all hardware and firmware items used in support of testing. If this information is available in another project document, it may be referenced instead of listing all equipment. The following is an example:

> *[Hardware Items] are to be used during testing, connected together in a client–server relationship through a TCP/IP-compatible network. For details of client and server hardware specifications, see the [PROJECT ABBREVIATION] Software Requirements Specifications (SRS). For integration and qualification level testing, a representative hardware set as specified in the SRS is to be used.*

Other Materials

This subsection should list all other materials required to support testing.

Proprietary Nature, Acquirer's Rights, and Licensing

Any of the issues associated with the potential proprietary nature of the software, acquirer's rights, or licensing should be addressed in this subsection. An example is provided:

Licensing of commercial software is one purchased copy for each PC it is to be used on, with the exception of [group software], which also requires licensing for the number of users/client systems. Some of the information in the dataset is covered by the Privacy Act and will have to be protected in some manner.

Installation, Testing, and Control

This subsection should address environmental requirements or actions required to support the installation of the application for testing. The following detail is provided.

[Company Name] will acquire the software and hardware needed as per the [PROJECT ABBREVIATION] Software Development Plan (SDP) in order to run [PROJECT ABBREVIATION] in accordance with contract directions and limitations.

[Company Name] will install the supporting software and test it per the procedures in the [PROJECT ABBREVIATION] Installation Software Test Description.

The [PROJECT ABBREVIATION] Change Enhancement Request (CER) tracking system will be used to determine eligibility for testing at any level. See the [PROJECT ABBREVIATION] Software Configuration Management Plan and [PROJECT ABBREVIATION] Software Quality Assurance Plan for details on CER forms and processes.

Participating Organizations

This subsection should identify all groups responsible for testing. They may also participate in completing the Test Plan Walk-through Checklist in Appendix C. An example is provided:

Test Sites. This section should provide a description of test sites. See below:

The planned installation level Beta test sites are listed as follows:
[Installation Beta Sites]

Software Items

This subsection should provide a complete list of all software items used in support of site testing, including versions. If the software is kept in an online repository, reference to the storage location may be cited instead of listing all items here. The following is an example:

Software used in the testing of [PROJECT ABBREVIATION]:
[Software Items]

For details of client and server software specifications see the [PRO-JECT ABBREVIATION] Software Requirements Specifications (SRS).

Hardware and Firmware Items

This subsection should provide a complete list of all hardware and firmware items used in support of site testing. If this information is available in another project document, it may be referenced instead of listing all equipment. The following is an example:

IBM compatible PCs are to be used during testing, connected together in a client–server relationship through a TCP/IP-compatible network. For details of client and server hardware specifications, see the [PROJECT ABBREVIATION] Software Requirements Specifications (SRS). For integration and qualification level, testing a representative hardware set as specified in the SRS is to be used.

Other Materials

This subsection should list all other materials required to support testing.

Proprietary Nature, Acquirer's Rights, and Licensing

Any of the issues associated with the potential proprietary nature of the software, acquirer's rights, or licensing should be addressed in this subsection. An example is provided:

Licensing of commercial software is one bought copy for each PC it is to be used on, with the exception of [group software], which also requires licensing for the number of users/client systems. Some of the information in the dataset is covered by the Privacy Act and will have to be protected in some manner.

Installation, Testing, and Control

This subsection should address environmental requirements or actions required to support the installation of the application for beta testing. The following detail is provided:

The customer will acquire the software and hardware needed as defined in the [PROJECT ABBREVIATION] SRS and the SOW in order to run [PROJECT ABBREVIATION].

[Company Name] will install the supporting software and test it per the procedures in the [PROJECT ABBREVIATION] Installation Software Test Description and as defined in the [PROJECT ABBREVIATION] SDP.

The CER tracking system will be used to determine eligibility for testing at any level. See the [PROJECT ABBREVIATION] Software Configu-

ration Management Plan and [PROJECT ABBREVIATION] Software Quality Assurance Plan for details on CER forms and processes.

Participating Organizations

This subsection should identify all groups responsible for testing. They may also participate in completing the Test Plan Walkthrough Checklist. An example is provided:

The participating organizations are [Company Name], [customer name], and organizations at the beta test site.

Test Levels

This subsection should describe the different levels of testing required in support of the development effort. See Appendix C, Test Design Specification, Test Case Specification, and Test Procedure Specification. The following is provided as an example:

Tests are to be performed at the module, integration, installation, and qualification levels prior to release for beta testing. Please refer to test design specifications #xx through xx.

Test Classes

This subsection should provide a description of all test classes. A test case specification should be associated with each of these classes. A summary of the validation method, data to be recorded, data analysis activity, assumptions, should be provided for each case. Example text is provided:

Test class # xx. Check for correct handling of erroneous inputs.

Test Objective. Check for proper handling of erroneous inputs: characters that are not valid for this field, too many characters, not enough characters, value too large, value too small, all selections for a selection list, no selections, all mouse buttons clicked or double clicked all over the client area of the item with focus.

Validation Methods Used—Test.

Recorded Data—User action or data entered, screen/view/dialog/control with focus, resulting action

Data Analysis—Was resulting action within general fault-handling capabilities defined in the [PROJECT ABBREVIATION] SRS and design in [PROJECT ABBREVIATION] SDD?

Assumptions and Constraints—None.

Additional examples are provided in support of test case development on the associated companion CD-ROM in *Software Test Plan.doc.*

General Test Conditions

This subsection should describe the anticipated baseline test environment. Refer to the information provided below:

> *A sample real dataset from an existing database is to be used during all tests. The sample real dataset will be controlled under the configuration management system. A copy of the controlled dataset is to be used in the performance of all testing.*

Test Progression

This subsection should provide a description of the progression of testing. A diagram is often helpful when attempting to describe the progression of testing (Figure 6-3). The information below is provided as an example:

> *The Qualification Testing is a qualification-level test verifying that all requirements have been met. The module and integration tests are performed as part of the Implementation phase as elements and modules are completed. All module and integration tests must be passed before performing Qualification Testing. All module tests must be passed before performing associated integration-level tests. The CER tracking system is used to determine eligibility for testing at a level. See the [PROJECT AB-BREVIATION] Software Configuration Management Plan and [PRO-*

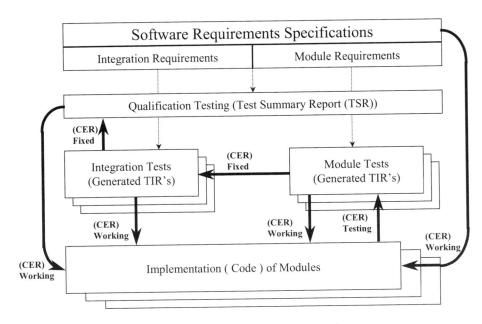

Figure 6-3. Software test progression.

JECT ABBREVIATION] Software Quality Assurance Plan for details on CER forms and processes.

Planned Tests

This section should provide a detailed description of the type of testing to be employed. The following is provided as an example:

A summary of testing is provided in Section 5. Additional information is provided in Appendix A, B, C, and D.

Qualification Test

This subsection should describe qualification test. An example is provided:

All of the requirement test items (refer to the Software Requirements Traceability Matrix) are to be tested as qualification-level tests. For details of the procedures and setup see the [PROJECT ABBREVIATION] Qualification Software Test Description. The resulting output of qualification test is the [PROJECT ABBREVIATION] Software Test Report (STR) and its attached Test Problem Reports (TPR). If qualification test is passed and its results accepted by the customer, the [PROJECT ABBREVIATION] software will be ready for beta release. Following Qualification Testing, the customer will review results. Signature of acceptance initiates product delivery and start of Installation beta test.

For the qualification-level tests the following classes of tests will be used:

> *Check for correct handling of erroneous inputs*
> *Check for maximum capacity*
> *User interaction behavior consistency*
> *Retrieving data*
> *Saving data*
> *Display screen and printing format consistency*
> *Check interactions between modules*
> *Measure time of reaction to user input*
> *Functional flow*

Integration Test

This subsection should describe the integration test. An example text is provided:

All of the modules to be integration tested (refer to Software Requirements Traceability Matrix) will be tested using integration-level test methodology. For details of the procedures and setup, see the [PROJECT ABBREVIATION] Integration Software Test Description. The resulting outputs of this test are Internal Test Reports (ITR) or System Integration Test Report (SITR). When these integration tests are all passed, the [PROJECT AB-BREVIATION] software will be ready for qualification-level testing.

For the integration-level tests the following classes of tests will be used:

> *User interaction behavior consistency*
> *Display screen and printing format consistency*
> *Check interactions between modules*
> *Measure time of reaction to user input*

Module Test

This section should describe module test. An example text is provided:

> *All of the modules to be tested (refer to the Software Requirements Trace-ability Matrix) are to be tested using defined module-level test methodol-ogy. For details regarding test procedures and setup, see the [PROJECT ABBREVIATION] Module Software Test Description. The resulting out-puts of this test are Internal Test Reports (ITR) or Unit Test Report (UTR). When all of the module tests for a module are passed, the module is ready for integration-level testing.*
>
> *For the module-level tests the following classes of tests will be used:*

> > *Check for correct handling of erroneous inputs*
> > *Check for maximum capacity*
> > *User interaction behavior consistency*
> > *Retrieving data*
> > *Saving data*
> > *Display screen and printing format consistency*
> > *Measure time of reaction to user input*

Installation Beta Test

This subsection should describe installation beta testing. An example text is provided:

> *Following qualification testing, the customer will review results. Signature of acceptance initiates product delivery and start of Installation beta test. Identified tests (refer to the Software Requirements Traceability Matrix) will be tested using defined installation-level Beta Test procedures. Some of these tests maybe documented in the Product Packaging Information in Appendix C. This test methodology will be ap-plied at each beta test site. For details regarding test procedures and setup see the [PROJECT ABBREVIATION] Installation Software Test Description. Outputs of these tests are Test Problem Reports (TPR). Following Installation Testing, the customer will review results. Signature of acceptance completes the [PROJECT ABBREVIATION] Version 1.0 project.*

For the installation level tests the following classes of tests will be used:

> *User interaction behavior consistency*
> *Retrieving data*
> *Saving data*
> *Display screen and printing format consistency*
> *Check interactions between modules*
> *Measure time of reaction to user input*
> *Functional flow*

Test Schedules

This section should include test milestones as identified in the software project schedule as well as any required deliverables associated with testing. Any additional milestones may be defined here as needed. If appropriate detail is provided in the Software Project Management Plan (SRMP), then this can be referenced. The following is provided as an example:

> *Test schedules are defined in the [PROJECT ABBREVIATION] Software Development Plan.*

Risk Management

This section should identify all high-risk assumptions of the test plan. The contingency plans for each should be described and the management of risk items should be addressed.

Requirements Traceability

This section should provide information regarding the traceability of testing to the requirements and design of the software. A matrix or database is useful when meeting this traceability requirement. The following example is provided:

> *Refer to [PROJECT ABBREVIATION] Software Requirements Specification, Appendix A for information regarding requirements traceability.*

Notes

This section should provide any additional information not previously covered in the test development plan.

System Test Plan

The following information is based on IEEE Std 829, IEEE Standard for Software Test Documentation [5], and IEEE Std 12207.0, Standard for Information Technology—Life Cycle Processes [39], and has been adapted to support ISO 9001 require-

Table 6-10. System test plan document outline

Title Page
Revision Page
Table of Contents
1. Introduction
2. Scope
    2.1   Identification and Purpose
    2.2   System Overview
    2.3   Definitions, Acronyms, and Abbreviations
3. Referenced Documents
4. System Test Objectives
5. Kinds of System Testing
    5.1   Functional Testing
    5.2   Performance Testing
    5.3   Reliability Testing
    5.4   Configuration Testing
    5.5   Availability Testing
    5.6   Portability Testing
    5.7   Security and Safety Testing
    5.8   System Usability Testing
    5.9   Internationalization Testing
    5.10 Operations Manual Testing
    5.11 Load Testing
    5.12 Stress Testing
    5.13 Robustness Testing
    5.14 Contention Testing
6. System Test Environment
    6.1   Development Test and Evaluation
        6.1.1   System and Software Items
        6.1.2   Hardware and Firmware Items
        6.1.3   Other Materials
        6.1.4   Proprietary Nature, Acquirer's Rights, and Licensing
        6.1.5   Installation, Testing, and Control
        6.1.6   Participating Organizations
    6.2   Test Site(s)
7. Test Identification
    7.1   General Information
        7.1.1   Test Levels
        7.1.2   Test Classes
        7.1.3   General Test Conditions
        7.1.4   Test Progression
    7.2   Planned Testing
        7.2.1   Module Test
        7.2.2   Integration Test
        7.2.3   System Test
        7.2.4   Qualification Test
        7.2.5   Installation Beta Test
8. Test Schedules

Table 6-10. Continued

9. Risk Management
10. Requirements Traceability
11. Notes
Appendix A. System Test Requirements Matrix
Appendix B. Module System Test Description
Appendix C. System Test Description
Appendix D. Qualification System Test Description
Appendix E. Installation System Test Description

ments. Additional information is provided in the document template, *System Test Plan.doc,* which is located on the companion CD-ROM. Table 6-10 is an outline of suggested document content.

The following provides section-by-section guidance in support of the creation of a system test plan. This guidance should be used to help define associated test requirements and should reflect the actual requirements of the implementing organization.

Introduction

This section should provide a brief introductory overview of the project and related testing activities described in this document.

Scope

Identification and Purpose. This subsection should provide information that uniquely identifies the system effort and this associated test plan. This can also be provided in the form of a unique test plan identifier. The following text provides an example:

> *This System test plan is to detail the testing planned for the [Project Name] ([Project Abbreviation]) Version [xx], Statement of Work [date], Task Order [to number], Contract No. [Contract #], and Amendments.*
>
> *The goal of [project acronym] development is to [goal]. This System will allow [purpose].*
>
> *Specifics regarding the implementation of these modules are identified in the [Project Abbreviation] System Requirements Specification (SysRS) with line item descriptions in the accompanying Requirements Traceability Matrix (RTM).*

System Overview. This section should provide a summary of all system items and system features to be tested. The need for each item and its history may be addressed here as well. References to associated project documents should be cited here. The following provides an example:

> *This document describes the System Test Plan (SysTP) for the [PROJECT ABBREVIATION] System. [PROJECT ABBREVIATION] documentation will include this SysTP, the [PROJECT ABBREVIATION] System Requirements Specification (SysRS), the System Requirements Traceability Matrix (RTM), the [PROJECT ABBREVIATION] Software Development Plan (SDP), the Software Design Document (SDD), the [PROJECT ABBREVIATION] User's Manual, the [PROJECT ABBREVIATION] System Administrator's Manual, and the [PROJECT ABBREVIATION] Data Dictionary.*
>
> *This SysTP describes the process to be used and the deliverables to be generated in the testing of the [PROJECT ABBREVIATION]. This plan, and the items defined herein, will comply with the procedures defined in the [PROJECT ABBREVIATION] Software Configuration Management (SCM) Plan and the [PROJECT ABBREVIATION] Software Quality Assurance (SQA) Plan. Any nonconformance to these plans will be documented as such in this SysTP.*
>
> *This document is based on the [PROJECT ABBREVIATION] System Test Plan template with tailoring appropriate to the processes associated with the creation of the [PROJECT ABBREVIATION]. The information contained in this SysTP has been created for [Customer Name] and is to be considered "For Official Use Only."*
>
> *Please refer to section "6. Schedule" in the SDP for information regarding documentation releases.*

Definitions, Acronyms, and Abbreviations

All relevant definitions, acronyms, and abbreviations should be included in this subsection.

Referenced Documents

This section should include all material referenced during the creation of the test plan.

System Test Objectives

System Test Objectives should include the validation of the application (i.e., to determine if it fulfills its system requirements specification), the identification of defects that are not efficiently identified during unit and integration testing, and the determination of the extent to which the system is ready for launch. This section should provide project status measures (e.g., percentage of test scripts successfully tested). A description of some of the kinds of testing may be found in Appendix C, Work Products, Example of System Testing.

System Test Environment

Development Test and Evaluation. Provide a high-level summary of all key players, facilities required, and site of test performance. The following is provided as an example:

> *System tests are to be performed at [company name], [site location]. All testing will be conducted in the development center [room number]. The following individuals must be in attendance: [provide list of performers and observers].*

System and Software Items. This subsection should provide a complete list of all system and software items used in support of testing, including versions. If the system is kept in an online repository, reference to the storage location may be cited instead of listing all items here. The following is an example:

> *System and software used in the testing of [PROJECT ABBREVIATION]:*
>   *[System and software List]*
>   *For details of client and server System specifications see the [PROJECT ABBREVIATION] System Requirements Specifications (SysRS).*

Hardware and Firmware Items. This subsection should provide a complete list of all hardware and firmware items used in support of testing. If this information is available in another project document, it may be referenced instead of listing all equipment. The following is an example:

> *[Hardware Items] are to be used during testing, connected together in a client–server relationship through a TCP/IP-compatible network. For details of client and server hardware specifications, see the [PROJECT ABBREVIATION] System Requirements Specifications (SysRS). For integration and qualification-level testing, a representative hardware set as specified in the SysRS is to be used.*

Other Materials. This subsection should list all other materials required to support testing.

Proprietary Nature, Acquirer's Rights, and Licensing. Any of the issues associated with the potential proprietary nature of the software, acquirer's rights, or licensing should be addressed in this section. An example is provided:

> *Licensing of commercial software is one purchased copy for each PC it is to be used on, with the exception of [group software], which also requires licensing for the number of users/client systems. Some of the information in the dataset is covered by the Privacy Act and will have to be protected in some manner.*

Installation, Testing, and Control. This subsection should address environment requirements or actions required to support the installation of the application for testing. The following detail is provided:

> *[Company Name] will acquire the software and hardware needed as per the [PROJECT ABBREVIATION] Software Development Plan (SDP) in order to run [PROJECT ABBREVIATION] in accordance with contract directions and limitations.*
>
> *[Company Name] will install the supporting software and test it per the procedures in the [PROJECT ABBREVIATION] Installation System Test Description.*
>
> *The [PROJECT ABBREVIATION] Change Enhancement Request (CER) tracking system will be used to determine eligibility for testing at any level. See the [PROJECT ABBREVIATION] Software Configuration Management Plan and [PROJECT ABBREVIATION] Software Quality Assurance Plan for details on CER forms and processes.*

Participating Organizations. This subsection should identify all groups responsible for testing. An example is provided:

> *The participating organizations are [Company Name], [customer name], and organizations at the beta test site.*

These organizations may participate by completing the Test Plan Walk-through Checklist as described in Appendix C, Work Products.

Test Site(s)

This subsection should provide a description of test sites. An example is provided below:

> *The planned installation level Beta test sites are listed as follows:*
> *[Installation Beta Sites]*

Test Identification—General Information

Test Levels. This subsection should describe the different levels of testing required in support of the development effort. See Appendix C, Test Design Specification, Test Case Specification, and Test Procedure Specification. The following is provided as an example:

> *Tests are to be performed at the module, integration, installation, and qualification levels prior to release for beta testing. Please refer to test design specifications # xx through xx.*

Test Classes. This subsection should provide a description of all test classes. A test case specification should be associated with each of these classes. A summary of the validation method, data to be recorded, data analysis activity, and assumptions, should be provided for each case. An example text is provided in Appendix C, Example Test Classes.

General Test Conditions. This subsection should describe the anticipated baseline test environment. Refer to the information provided below:

> *A sample real dataset from an existing database is to be used during all tests. The sample real dataset will be controlled under the configuration management system. A copy of the controlled data set is to be used in the performance of all testing.*

Test Progression. This subsection should provide a description of the progression of testing. A diagram is often helpful when attempting to describe the progression of testing (Figure 6-4). The information below is provided as an example:

> *The Qualification Testing is a qualification-level test verifying that all requirements have been met. The module and integration tests are performed as part of the Implementation phase as elements and modules are completed. All module and integration tests must be passed before per-*

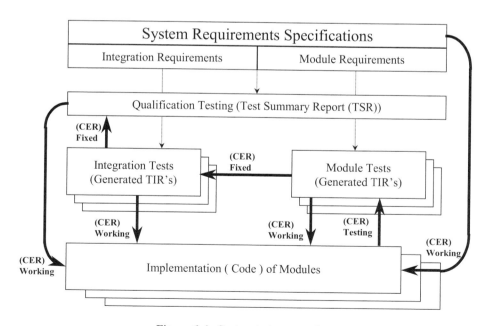

Figure 6-4. System test progression.

*forming System Testing. All module tests must be passed before perform-
ing associated integration-level tests. The CER tracking system is used to
determine eligibility for testing at a level. See the [PROJECT ABBREVI-
ATION] Software Configuration Management Plan and [PROJECT AB-
BREVIATION] Software Quality Assurance Plan for details on CER
forms and processes.*

Planned Testing

This subsection should provide a detailed description of the type of testing to be
employed. Examples text follows:

*Module Test. All of the modules to be tested (refer to the Software Re-
quirements Traceability Matrix) are to be tested using defined module-
level test methodology. For details regarding test procedures and setup,
see the [PROJECT ABBREVIATION] Module Software Test Description.
The resulting outputs of this test are Internal Test Reports (ITR) or Unit
Test Report (UTR). When all of the modules test are passed for a module,
the module is ready for integration testing.*

For the module-level tests, the following classes of tests will be used:

> *Check for correct handling of erroneous inputs
> Check for maximum capacity
> User interaction behavior consistency
> Retrieving data
> Saving data
> Display screen and printing format consistency
> Measure time of reaction to user input*

*Integration Test. All of the modules to be integration tested (refer to Sys-
tem Requirements Traceability Matrix) will be tested using integration-
level test methodology. For details of the procedures and setup, see the
[PROJECT ABBREVIATION] Integration System Test Description. The
resulting outputs of this test are Internal Test Reports (ITR) or System Integration Test Report (SITR).*

*When these integration tests are all passed, the [PROJECT ABBREVI-
ATION] System will be ready for qualification-level testing.*

For the integration level tests the following classes of tests will be used:

> *User interaction behavior consistency
> Display screen and printing format consistency
> Check interactions between modules
> Measure time of reaction to user input*

*System Test. All of the requirement test items (refer to the System Re-
quirements Traceability Matrix) are to be tested as system-level tests. For
details of the procedures and setup, see the [PROJECT ABBREVIA-
TION] System Test Plan Description. The resulting output of system test*

is the [PROJECT ABBREVIATION] System Test Report (SysTR) and its attached Test Problem Reports (TPR). If system test is passed and its results accepted by the [STAKEHOLDER] the [PROJECT ABBREVIATION] System will be ready for Qualification Test. Following Qualification Testing, the customer will review results. Signature of acceptance initiates product delivery and start of Installation beta test.

Qualification Test. All of the requirement test items (refer to the System Requirements Traceability Matrix) are to be tested as qualification-level tests. For details of the procedures and setup, see the [PROJECT ABBREVIATION] Qualification System Test Description. The resulting output of qualification test is the [PROJECT ABBREVIATION] System Test Report (SysTR) and its attached Test Problem Reports (TPR). If qualification test is passed and its results accepted by the customer, the [PROJECT ABBREVIATION] System will be ready for beta release. Following Qualification Testing, the customer will review results. Signature of acceptance initiates product delivery and start of Installation beta test.

For the qualification level tests the following classes of tests will be used:

> *Check for correct handling of erroneous inputs*
> *Check for maximum capacity*
> *User interaction behavior consistency*
> *Retrieving data*
> *Saving data*
> *Display screen and printing format consistency*
> *Check interactions between modules*
> *Measure time of reaction to user input*
> *Functional flow*

Installation Beta Test. Following qualification testing the customer will review results. Signature of acceptance initiates product delivery and start of Installation beta test. Identified tests (refer to the System Requirements Traceability Matrix) will be tested using defined installation-level Beta Test procedures. Some of these tests maybe documented in the Product Packaging Information in Appendix C. This test methodology will be applied at each beta test site. For details regarding test procedures and setup see the [PROJECT ABBREVIATION] Installation System Test Description. Outputs of these tests are Test Problem Reports (TPR). Following Installation Testing, the customer will review results. Signature of acceptance completes the [PROJECT ABBREVIATION] Version 1.0 project.

For the installation-level tests, the following classes of tests will be used:

> *User interaction behavior consistency*
> *Retrieving data*

Saving data
Display screen and printing format consistency
Check interactions between modules
Measure time of reaction to user input
Functional flow

Test Schedules

This section should include test milestones as identified in the system project schedule as well as any required deliverables associated with testing. Any additional milestones may be defined here as needed. If appropriate detail is provided in the software project management plan (SRMP), then this can be referenced. The following is provided as an example:

Test schedules are defined in the [PROJECT ABBREVIATION] Software Development Plan.

Risk Management

This section should identify all high-risk assumptions of the test plan. The contingency plans for each should be described and the management of risk items should be addressed.

Requirements Traceability

This section should provide information regarding the traceability of testing to the requirements and design of the system. A matrix or database is useful when meeting this traceability requirement. The following example is provided:

Refer to [PROJECT ABBREVIATION] System Requirements Specification, Appendix A for information regarding requirements traceability.

Note. This section should provide any additional information not previously covered in the test development plan.

System Integration Test Plan

The following information is based on IEEE Std 829, IEEE Standard for Software Test Documentation [5], and IEEE Std 12207.0, Standard for Information Technology—Life Cycle Processes [39], and has been adapted to support ISO 9001 requirements. Additional information is provided in the document template, *System Integration Test Plan.doc,* which is located on the companion CD-ROM. Table 6-10 is an outline of suggested document content.

The following provides section-by-section guidance in support of the creation of a system integration test plan. This guidance should be used to help define associat-

Table 6-10. System integration test plan document outline

Title Page
Revision Page
Table of Contents
  1. Introduction
  2. Scope
      2.1   Identification and Purpose
      2.2   System Overview
      2.3   Definitions, Acronyms, and Abbreviations
  3. Referenced Documents
  4. System Integration Test Objectives
  5. System Integration Testing Kinds and Approaches
      5.1   Software Integration Testing
      5.2   COTS Integration Testing
      5.3   Database Integration Testing
      5.4   Hardware Integration Testing
      5.5   Prototype Usability Testing
      5.6   Approaches
  6. System Integration Test Environment
      6.1   Development Test and Evaluation
            6.1.1   System and Software Items
            6.1.2   Hardware and Firmware
            6.1.3   Other Materials
            6.1.4   Proprietary Nature, Acquirer's Rights, and Licensing
            6.1.5   Participating Organizations
      6.2   Test Sites
  7. Test Identification
      7.2   General Information
            7.1.1   Test Levels
            7.1.2   Test Classes
            7.1.3   General Test Conditions
            7.1.4   Test Progression
      7.2   Planned Testing
            7.2.1   Integration Test
  8. Test Schedules
  9. Risk Management
 10. Requirements Traceability
 11. Notes
Appendix A. System Test Requirements Matrix
Appendix B. System Integration Test Description

ed testing requirements and should reflect the actual requirements of the implementing organization.

Introduction

This section should provide a brief introductory overview of the project and related testing activities described in this document.

Scope

Identification and Purpose. This subsection should provide information that uniquely identifies the system effort and the associated test plan. This can also be provided in the form of a unique test plan identifier. The following text provides an example:

> *This System Integration Test Plan is to detail the testing planned for the [Project Name] ([Project Abbreviation]) Version [xx], Statement of Work [date], Task Order [to number], Contract No. [Contract #], and Amendments.*
>
> *The goal of [project acronym] development is to [goal]. This System will allow [purpose].*
>
> *Specifics regarding the implementation of these modules are identified in the [Project Abbreviation] System Requirements Specification (SysRS) with line item descriptions in the accompanying Requirements Traceability Matrix (RTM) and references to the Interface Control Documentation (ICD).*

System Overview. This section should provide a summary of all system items and system features to be tested. The need for each item and its history may be addressed here as well. References to associated project documents should be cited here. The following provides an example:

> *This document describes the System Integration Test Plan (SysITP) for the [PROJECT ABBREVIATION] System. [PROJECT ABBREVIATION] documentation will include this SysITP, the [PROJECT ABBREVIATION] System Requirements Specification (SysRS), the System Requirements Traceability Matrix (RTM), the Interface Control Documentation (ICD), the [PROJECT ABBREVIATION] Software Development Plan (SDP), the Software Design Document (SDD), the [PROJECT ABBREVIATION] User's Manual, the [PROJECT ABBREVIATION] System Administrator's Manual, and the [PROJECT ABBREVIATION] Data Dictionary.*
>
> *This SysITP describes the process to be used and the deliverables to be generated in the testing of the [PROJECT ABBREVIATION]. This plan, and the items defined herein, will comply with the procedures defined in the [PROJECT ABBREVIATION] Software Configuration Management*

(SCM) Plan and the [PROJECT ABBREVIATION] Software Quality Assurance (SQA) Plan. Any nonconformance to these plans will be documented as such in this SysITP.

This document is based on the [PROJECT ABBREVIATION] System Integration Test Plan template with tailoring appropriate to the processes associated with the creation of the [PROJECT ABBREVIATION]. The information contained in this SysITP has been created for [Customer Name] and is to be considered "For Official Use Only."

Please refer to section "6. Schedule" in the SDP for information regarding documentation releases.

Definitions, Acronyms, and Abbreviations

All relevant definitions, acronyms, and abbreviations should be included in this subsection.

Referenced Documents

This section should include all material referenced during the creation of the test plan

System Integration Test Objectives

Test objectives should include the verification of the product integration (i.e., to determine if it fulfills its system requirements specification and architectural design), the identification of defects that are not efficiently identified during unit testing, and the determination of the extent to which the subsystems/system is ready for system test. This section should provide project status measures (e.g., percentage of test scripts successfully tested).

System Integration Testing Kinds and Approaches

Software Integration Testing. Incremental integration testing of two or more integrated software components on a single platform or multiple platforms is conducted to produce failures caused by interface defects. This section should provide references to the SysRS, design documentation, and interface control document to show a complete list of subsystem or system modules and resulting test strategy.

COTS Integration Testing. Incremental integration testing of multiple commercial-off-the-shelf (COTS) software components is conducted to determine if they are not interoperable (i.e., if they contain any interface defects). This subsection should provide references to the SysRS, design documentation, and interface control document to show a complete list of subsystem or system modules and resulting test strategy.

Database Integration Testing. Incremental integration testing of two or more integrated software components is conducted to determine if the application software components interface properly with the database(s). This subsection should provide references to the SysRS, design documentation, and Interface control document to show a complete list of subsystem or system modules and resulting test strategy.

Hardware Integration Testing. Incremental integration testing of two or more integrated hardware components in a single environment is conducted to produce failures caused by interface defects. This subsection should provide references to the SysRS, design documentation, and Interface control document to show a complete list of subsystem or system modules and resulting test strategy.

Prototype Usability Testing. Incremental integration testing of a user interface prototype against its usability requirements is conducted to determine if it contains any usability defects. This subsection should provide references to the SysRS, design documentation, and user manual to show a complete list of subsystem or system modules and resulting test strategy.

Approaches. Top-down testing consists of high-level components of a system are integrated and tested before their design and implementation has been completed. Bottom-up testing consists of low-level components integrated and tested before the high-level components have been developed. Object-oriented testing consists of use-case or scenario-based testing, Thread testing, and/or object interaction testing. Interface testing consists of modules or subsystems that are integrated to create large systems.

System Integration Test Environment

Development Test and Evaluation. Provide a high-level summary of all key players, facilities required, and site of test performance. The following is provided as an example:

> *System Integration Tests are to be performed at [company name], [site location]. All integration testing will be conducted in the development center [room number]. The following individuals must be in attendance: [provide list of performers and observers].*

System and Software Items. This subsection should provide a complete list of all system and software items used in support of integration testing, including versions. If the system is kept in an online repository, reference to the storage location may be cited instead of listing all items here. The following is an example:

> *System and software used in the integration testing of [PROJECT AB-BREVIATION]:*

[System and software List]
For details of client and server System specifications see the [PRO-JECT ABBREVIATION] System Requirements Specifications (SysRS).

Hardware and Firmware Items. This subsection should provide a complete list of all hardware and firmware items used in support of integration testing. If this information is available in another project document, it may be referenced instead of listing all equipment. The following is an example:

[Hardware Items] are to be used during integration testing, connected together in a client–server relationship through a TCP/IP-compatible network. For details of client and server hardware specifications, see the [PROJECT ABBREVIATION] System Requirements Specifications (SysRS). For integration testing, a representative hardware set as specified in the SysRS is to be used.

Other Materials. This section should list all other materials required to support integration testing.

Proprietary Nature, Acquirer's Rights, and Licensing. Any of the issues associated with the potential proprietary nature of the software, acquirer's rights, or licensing should be addressed in this subsection. An example is provided:

Licensing of commercial software is one purchased copy for each PC it is to be used on, with the exception of [group software], which also requires licensing for the number of users/client systems. Some of the information in the dataset is covered by the Privacy Act and will have to be protected in some manner.

Participating Organizations. This subsection should identify all groups responsible for integration testing. An example is provided:

The participating organizations are [Company Name], [customer name], and organizations at the beta test site.

They may participate in completing the Test Plan Walk-through Checklist. An example is identified in Appendix C, Work Products, of this text.

Test Site(s)

This subsection should provide a description of test sites. An example is given below:

The planned integration test sites are listed as follows:
[Integration Sites]

Test Identification—General Information

Test Levels. This subsection should describe the different levels of testing required in support of the development effort. See Appendix C for Test Design Specification, Test Case Specification, and Test Procedure Specification. The following is provided as an example:

> *Tests are to be performed at the integration level prior to release for system testing. Please refer to test design specifications # xx through xx.*

Test Classes. This subsection should provide a description of all test classes. A test case specification should be associated with each of these classes. A summary of the validation method, data to be recorded, data analysis activity, and assumptions should be provided for each case.

General Test Conditions. This subsection should describe the anticipated baseline test environment. Refer to the information provided below:

> *A sample real dataset from an existing database is to be used during all tests. The sample real dataset will be controlled under the configuration management system. A copy of the controlled dataset is to be used in the performance of all testing.*

Test Progression. This subsection should provide a description of the progression of testing. A diagram is often helpful when attempting to describe the progression of testing. The information below is provided as an example:

> *The System Integration Testing is testing that verifies that all subsystems or modules work together. The module tests are performed prior to or during the Implementation phase as elements and modules are completed. All module and integration tests must be passed before performing System Testing. The Change Enhancement Request (CER) tracking system is used to determine eligibility for testing at a level. See the [PROJECT ABBREVIATION] Software Configuration Management Plan and [PROJECT ABBREVIATION] Software Quality Assurance Plan for details on CER forms and processes.*

Planned Testing

This subsection should provide a detailed description of the type of testing to be employed. The following is provided as an example:

> *A summary of testing is provided in Section xxx. Additional information is provided in Appendixes A and B.*

Integration Test. An example of this subsection follows:

> *All of the modules to be integration tested (refer to System Requirements Traceability Matrix) will be tested using integration-level test methodology. For details of the procedures and setup, see the [PROJECT ABBRE-VIATION] System Integration Test Description. The resulting outputs of this test are Internal Test Reports (ITR) or System Integration Test Report (SITR). When these integration tests are all passed, the [PROJECT AB-BREVIATION] System will be ready for system-level testing.*
>
> *For the integration-level tests, the following classes of tests will be used:*
>
> > *User interaction behavior consistency*
> > *Display screen and printing format consistency*
> > *Check interactions between modules*
> > *Measure time of reaction to user input*

Test Schedules

This section should include test milestones as identified in the system project schedule as well as any required deliverables associated with testing. Any additional milestones may be defined here as needed. If appropriate detail is provided in the software project management plan (SRMP), then this can be referenced. The following is provided as an example:

> *Test schedules are defined in the [PROJECT ABBREVIATION] Software Development Plan.*

Risk Management

This section should identify all high-risk assumptions of the test plan. The contingency plans for each should be described and the management of risk items should be addressed.

Requirements Traceability

This section should provide information regarding the traceability of testing to the requirements and design of the system. A matrix, or database, is useful when meeting this traceability requirement. The following is provided as an example:

> *Refer to [PROJECT ABBREVIATION] System Requirements Specification, Appendix A for information regarding requirements traceability.*

Note. This section should provide any additional information not previously covered in the test development plan.

Appendix A. System Test Requirements Matrix

Refer to SysRS, Section 6, and attach test information as another column.

Appendix B. System Integration Test Description

This should contain a detailed description of each test.

JOINT REVIEW

The joint review process evaluates the status and products of an activity of a project as appropriate. Joint reviews are carried out at both project management and technical levels and are held throughout the life of the contract. Joint reviews may include the customer and/or the supplier. Examples are project management reviews and technical reviews. Throughout the development process, the technical reviews complete the activities of:

- Software requirements analysis
- Software architectural design
- Software detailed design
- Software integration
- Software acceptance support

Table 6-11 provides a list of the joint review process objectives.

ISO 9001 Goals

ISO 9001 describes requirements in support of the IEEE 12207 Joint Review Process. These technical and management reviews are planned in Clause 7.1 across the software life cycle, starting with software requirements and continuing through customer communications, planning, and reviews. The five requirement clauses follow.

> Clause 7.1, Planning of Product Realization, requires a life cycle model to describe and relate all the processes, which must be planned and developed. The planning must determine the inspection activities, such as the project manage-

Table 6-11. Joint review process objectives

a) Evaluate the status and products of an activity of a process through joint review activities between the parties to a contract.

b) Establish mechanisms to ensure that action items raised are recorded for action.

ment review to confirm the completion criteria for starting and ending each project stage.

Clause 7.2.2, Review of Requirements Related to the Product, requires the software product requirements review before committing to supply the product to the customer. This becomes a joint review when the customer or its advocate participates.

Clause 7.2.3, Customer Communication, determines and implements effective arrangements for communicating with customers during the product realization phases. Such arrangements during development can include joint reviews.

Clause 7.3.4, Design and Development Review, requires systematic reviews to be performed at suitable stages in accordance with planned arrangements. Such arrangements during development can include joint reviews, such as technical and project management reviews.

Clause 7.3.1, Design and Development Planning, requires planning and controlling the software product design and development. The interfaces between the different involved groups must be managed to ensure effective communication and the clear assignment of responsibility. These interfaces participate in joint reviews for planning.

Technical Reviews

The information provided here is based upon the recommendations provided in IEEE Std 1028, IEEE Standard for Software Reviews. Technical reviews are an effective way to detect and identify problems early in the software development process. The criteria for technical reviews are presented here to help organizations define their review practices.

Introduction

Technical reviews are an effective way to evaluate a software product by a team of qualified personnel to identify discrepancies from specifications and standards. Although technical reviews identify anomalies, they may also provide the recommendation and examination of various alternatives. The examination need not address all aspects of the product and may only focus on a selected aspect of a software product.

Software requirements specifications, design descriptions, test and user documentation, installation and maintenance procedures, and build processes are all examples of items subject to technical reviews.

Responsibilities

The roles in support of a technical review are:

Decision maker. The decision maker is the individual requesting the review and who determines whether objectives have been met.

Review leader. The review leader is responsible for the review and must perform all administrative tasks (to include summary reporting), ensure that the review is conducted in an orderly manner, and that the review meets its objectives.

Recorder. The recorder is responsible for the documentation of all anomalies, action items, decisions, and recommendations made by the review team.

Technical staff. The technical staff should actively participate in the review and evaluation of the software product.

The following roles are optional and may also be established for the technical review:

Management staff. Management staff may participate in the technical review for the purpose of identifying issues that will require management resolution.

Customer or user representative. The role of the customer or user representative should be determined by the review leader prior to the review.

Input

Input to the technical review should include a statement of objectives, the software product being examined, existing anomalies and review reports, review procedures, and any standard against which the product is to be examined. Anomaly categories should be defined and be available during the technical review. For additional information in support of the categorization of software product anomalies, refer to IEEE Std 1044 [13].

Authorization

All technical reviews should be defined in the SPMP. The plan should describe the review schedule and all allocation of resources. In addition to those technical reviews required by the SPMP, other technical reviews may be scheduled.

Preconditions

A technical review should be conducted only when the objectives have been established and the required review inputs are available.

Procedures

Management Preparation. Managers are responsible for ensuring that all reviews are planned, that all team members are trained and knowledgeable, and that adequate resources are provided. They are also responsible for ensuring that all review procedures are followed.

Planning the Review. The review leader is responsible for the identification of the review team and their assignment of responsibility. The leader should schedule and announce the meeting, prepare participants for the review by providing them with the required material, and collecting all comments.

Overview of Review Procedures. The team should be presented with an overview of the review procedures. This overview may occur as a part of the review meeting or as a separate meeting.

Overview of the Software Product. The team should receive an overview of the software product. This overview may occur either as a part of the review meeting or as a separate meeting.

Preparation

All team members are responsible for reviewing the product prior to the review meeting. All anomalies identified during this prereview process should be presented to the review leader. Prior to the review meeting, the leader should classify all anomalies and forward these to the author of the software product for disposition.

The review leader is also responsible for the collection of all individual preparation times to determine the total preparation time associated with the review.

Examination

The review meeting should have a defined agenda that should be based upon the premeeting anomaly summary. Based upon the information presented, the team should determine whether the product is suitable for its intended use, whether it conforms to appropriate standards, and whether is ready for the next project activity. All anomalies should be identified and documented.

Rework/Follow-up

The review leader is responsible for verifying that the action items assigned in the meeting are closed.

Exit Criteria

A technical review is considered complete when all follow-up activities have been completed and the review report has been published.

Output

The output from the technical review should consist of the project being reviewed, a list of the review team members, a description of the review objectives, and a list of resolved and unresolved software product anomalies. The output should also include a list of management issues, all action items and their status, and any recommendations made by the review team.

Management Reviews

The information provided here is based upon the recommendations provided in IEEE Std 1028, IEEE Standard for Software Reviews. Management reviews are an effective way to detect and identify problems early in the software development process. The criteria for management reviews are presented here to help organizations define their review practices. It is important to remember that management reviews are not only about the specific review of products, but also may cover aspects of the software process.

Introduction

The purpose of a management review is to monitor project progress and to determine the status according to documented plans and schedules. Reviews can also be used to evaluate the effectiveness of the management approaches. The effective use of management reviews can support decisions about corrective action, resource allocation, or changes in scope.

Management reviews may not address all aspects of a given project during a single review but may require several review cycles to completely evaluate a software product. Examples of software products subject to management review include:

a) Anomaly reports
b) Audit reports
c) Back-up and recovery plans
d) Contingency plans
e) Customer complaints
f) Disaster plans
g) Hardware performance plans
h) Installation plans
i) Maintenance plans
j) Procurement and contracting methods
k) Progress reports
l) Risk management plans
m) Software configuration management plans
n) Software project management plans
o) Software quality assurance plans
p) Software safety plans
q) Software verification and validation plans
r) Technical review reports

s)   Software product analyses

t)   Verification and validation reports

Responsibilities

The management personnel associated with a given project should carry out management reviews. The individuals qualified to evaluate the software product should perform all management reviews:

Decision maker. The review is conducted for the decision maker and it is up to this individual to determine whether the objectives of the review have been met.

Review leader. The review leader is responsible for all administrative tasks required in support of the review. The review leader should ensure that all planning and preparation have been completed, that all objectives are established and met, and that the review is conducted and review outputs are published.

Recorder. The recorder is responsible for the documentation of all anomalies, action items, decisions, and recommendations made by the review team.

Management staff. The management staff is responsible for carrying out the review. They are responsible for active participation in the review.

Technical staff. The technical staff are responsible for providing the information required in support of the review.

Customer or user representative. The role of the customer or user representative should be determined by the review leader prior to the review.

Input

Input to the management review should include the following:

a) A statement of review objectives

b) The software product being evaluated

c) The software project management plan

d) Project status

e) A current anomalies list

f) Documented review procedures

g) Status of resources, including finance, as appropriate

h) Relevant review reports

i) Any associated standards

j) Anomaly categories

Authorization

The requirement for conducting management reviews should initially be established in the project planning documents. The completion of a software product or completion of a scheduled activity may initiate a management review. In addition to those management reviews required by a specific plan, other management reviews may be announced and held.

Preconditions

A management review should be conducted only when the objectives for the review have been established and all required inputs are available.

Management Preparation

Managers should ensure that the review is performed as planned and that appropriate time and resources are allocated in support of the review process. They should ensure that all participants are technically qualified and have received training and orientation in support of the review. They are responsible for ensuring that all reviews are carried out as planned and that any resulting action items are completed.

Planning the Review

The review leader is responsible for identifying the review team and the assignment of specific responsibilities. He/she is responsible for scheduling the meeting and for the distribution of all materials required in support of review preparation activities. Prereview comments should be collected by the review leader, classified, and be presented to the author prior to the review.

Overview of Review Procedures

A qualified person should present an overview of the review procedures for the review team if requested by the review leader.

Preparation

Each review team member should review the software product and any other review inputs prior to the review meeting. All anomalies detected during this examination should be documented and sent to the review leader.

Examination

The management review should consist of one or more meetings of the review team. The meetings should review the objectives of the review, evaluate the software product and product status, review all items identified prior to the review meeting, and generate a list of action items to include associated risk. The meeting should be documented. This documentation should include any risk issues that are critical to the success of the project. It should provide a confirmation of the software requirements, list action items, and identify other issues that should be addressed.

Rework/Follow-up

The review leader is responsible for ensuring that all action items assigned in the meeting are closed.

Exit Criteria

The management review is complete when the activities listed and all required outputs and follow-up activities are finished.

Output

The output from the management review should identify:

- a) The project being reviewed
- b) The review team members
- c) Review objectives
- d) Software product reviewed
- e) Specific inputs to the review
- f) All action items
- g) A list of all anomalies

AUDIT

The audit process determines the degree of organizational and project compliance with the processes, requirements, plans, and contract. This process is employed by one party (auditing party) who audits the software products or activities of another party (audited party). The audit produces a list of detected issues or problems, which are recorded and entered into the problem resolution process. Table 6-12 provides a list of the audit process objectives.

Table 6-12. Audit process objectives

---

a) Determine compliance with requirements, plans, and contract, as appropriate.
b) Arrange the conduct of audits of work products or process performance by a qualified independent party, as specified in the plans.
c) Conduct follow-up audits to assess corrective action(s), closure, and root cause actions.

ISO 9001 Goals

ISO 9001 describes requirements in support of the IEEE 12207 audit process. The key requirement is Clause 8.2.2, Internal Audit. This requires conducting periodic internal audits to determine if the quality management system conforms to planned arrangements and is effectively implemented and maintained. Management must ensure that actions are taken without undue delay to eliminate detected nonconformities and their causes. Follow-up actions must verify implementation of the action and report the results. The audit findings are inputs to Clause 5.6.2, Management Review Inputs, and also to Clause 8.5.1, Continual Improvement, according to which the organization must continually improve the effectiveness of the quality management system.

One variation of audits is the configuration audit to support Clause 7.3.6, Design and Development Validation, to confirm that the resulting product is capable of meeting the requirements for its specified application or intended use, where known.

Another audit variation is the supplier audit to support Clause 7.4.1, Purchasing Process, for the organization to evaluate and select suppliers based on their ability to supply a product in accordance with the requirements.

Audits

The information provided here is based upon the recommendations provided in IEEE Std 1028, IEEE Standard for Software Reviews. Audits are effective ways to detect and identify problems in the software development process and to ensure that practices and procedures are being implemented as expected. Audits can also be performed by comparing what people are doing against established plans and procedures. The criteria for audits are presented here to help organizations define their review practices.

Introduction

A software audit provides an independent evaluation of conformance. Examples of software products subject to audit include the following:

a) Back-up and recovery plans
b) Contingency plans
c) Contracts
d) Customer complaints
e) Disaster plans
f) Hardware performance plans
g) Installation plans and procedures
h) Maintenance plans
i) Management review reports

j) Operations and user manuals
k) Procurement and contracting methods
l) Reports and data
m) Risk management plans
n) Software configuration management plans
o) Software design descriptions
p) Source code
q) Unit development folders
r) Software project management plans
s) Software quality assurance plans
t) Software requirements specifications
u) Software safety plans
v) Software test documentation
w) Software user documentation
x) Software verification and validation plans

Responsibilities

The following roles should be established for an audit:

Lead auditor. The lead auditor is responsible for all administrative tasks, assembly of the audit team and their management, and for ensuring that the audit meets its objectives. The lead auditor should prepare a plan for the audit and summary activities in an audit report. The lead auditor should be free from bias and influence that could reduce any ability to make independent, objective evaluations.

Recorder. The recorder should document all anomalies, action items, decisions, and recommendations made by the audit team.

Auditor. The auditors should examine all products as described in the audit plan. They should record their observations and recommend corrective actions. All auditors should be free from bias and influences that could reduce their ability to make independent, objective evaluations.

Initiator. The initiator determines the need, focus, and purpose of the audit. The initiator determines the members of the audit team and reviews the audit plan and audit report.

Audited organization. The audited organization should provide a liaison to the auditors and is responsible for providing all information requested by the auditors. When the audit is completed, the audited organization should implement corrective actions and recommendations.

Input. Inputs to the audit should be listed in the audit plan and should include the purpose and scope of the audit, a list of products to be audited, and audit evaluation criteria.

Authorization. An initiator decides upon the need for an audit. A project milestone or a nonroutine event, such as the suspicion or discovery of a major nonconformance, may drive this decision. The initiator selects an auditing group and provides the auditors with all supporting information. The lead auditor produces an audit plan and the auditors prepare for the audit.

Preconditions. An audit should only be conducted after the audit has proper authorization, the audit objectives have been established, and all audit items are available.

Procedures

Management Preparation. Managers are responsible for ensuring that the appropriate time and resources have been planned in support of the audit activities. Managers should also ensure that adequate training and orientation on audit procedures is provided.

Planning the Audit. The audit plan should describe the purpose and scope of the audit; the audited organization, including location and management; and a list of software products to be audited. The plan should describe all evaluation criteria, auditor's responsibilities, examination activities, resource requirements and schedule, and required reporting.

The initiator should approve the audit plan but the plan should be flexible enough to allow for changes based on information gathered during the audit, subject to approval by the initiator.

Opening Meeting. An opening meeting between the audit team and audited organization should be scheduled to occur at the beginning of the examination activity of the audit. This meeting should cover the purpose and scope of the audit, all software products being audited, all audit procedures and outputs, audit requirements, and audit schedule.

Preparation

The initiator should notify the audited organization's management in writing before the audit is performed, except for unannounced audits. The purpose of notification is to ensure that the people and material to be examined in the audit are available.

Auditors should prepare for the audit by reviewing the audit plan and any information available about the audited organization and the products to be audited. The lead auditor should prepare for the team orientation and any necessary training. The lead auditor should also ensure that facilities are available for the audit interviews and that all materials are available.

Examination

The examination should consist of evidence collection and analysis with respect to the audit criteria as describe in the audit plan. A closing meeting between the auditors and audited organization should be conducted, and then an audit report should be prepared.

Evidence Collection

The auditors should collect evidence of conformance and nonconformance by interviewing audited organization staff, examining documents, and witnessing processes. The auditors should attempt all the examination activities defined in the audit plan and undertake additional activities if they consider them to be required to determine conformance. Auditors should document all observations of conformance or nonconformance. These observations should be categorized as major or minor. An observation should be classified as major if the nonconformity will likely have a significant effect on product quality, project cost, or project schedule. All observations should be discussed with the audited organization before the closing audit meeting.

Closing Meeting

The lead auditor should convene a closing meeting with the audited organization's management. The closing meeting should review the progress of the audit, all audit observations, and preliminary conclusions and recommendations. Agreements should be reached during the closing audit meeting and must be completed before the audit report is finalized.

Reporting

The lead auditor is responsible for the preparation of the audit report. The audit report should be prepared as quickly as possible following the audit. Any communication between auditors and the audited organization made between the closing meeting and the issue of the report should pass through the lead auditor. The lead auditor should send the audit report to the initiator who will distribute the audit report within the audited organization.

Follow-up

The initiator and audited organization should determine what corrective action may be required and the type of corrective action to be performed.

Exit Criteria

An audit is complete when the audit report has been delivered to the initiator and all audit recommendations have been performed.

Output

The output of the audit is the audit report. The audit report should describe the purpose and scope of the audit; the audited organization, including participants; identification of all products included in the audit; and recommendations. All criteria (e.g., standards and procedures) in support of the audit should be identified in the plan. A summary of all audit activities should be provided. All classified observations should be included, as well as a summary and interpretation of the findings. A schedule and list of all follow-up activities should be described.

Software Measurement and Measures Plan

The following provides a suggested format for a project-level measurement and measures* plan. This plan should contain a description of all measurement and analysis used in support of an identified software effort.

The modification of the recommended software measurement and measures plan table of contents to support the goals of ISO 9001 more directly is shown in Table 6-13. Additional information in support of related work products is provided in Appendix C, Work Products. These include a list of measures for reliable software and the measurement information model as described in ISO/IEC 15939.

Software Measurement and Measures Plan Document Guidance

The following provides section-by-section guidance in support of the creation of a software measurement and measures plan (MMP). The MMP should be considered

---

*Measures is a problematic term because it has no counterpart in generally accepted metrology. Generally, it is used in software engineering to mean a measurement coupled with a judgmental threshold of acceptability.

Table 6-13. Software measurement and measurements plan document outline

Title Page
Revision Page
Table of Contents
1. Introduction
    1.1 Purpose
    1.2 Scope
    1.3 Definitions, Acronyms, and Abbreviations
    1.4 References
2. Measurements Process Management
    2.1 Responsibilities
    2.2 Life Cycle Reporting
    2.3 General Information
3. Measurement and Measures

to be a living document. The MMP should change to reflect any process improvement or changes in procedures relating to software measurement and measures activities. This guidance should be used to help define a measurement and measures process and should reflect the actual processes and procedures of the implementing organization. Additional information is provided in the document template, *Software Measurement and Measures Plan.doc,* which is located on the companion CD-ROM.

Introduction

The introduction is not required, but can be used to state the goals for measurement activities. The following information is provided as an example introduction:

> *The goal of the [Project Name] ([Project Abbreviation]) software development effort is to [Project Goal]. This software will provide a method for [Customer Name] to efficiently meet their customer's requirements.*
>
> *This Software Measurement and Measures (SMM) Plan has been developed to ensure that [Project Abbreviation] software products conform to established technical and contractual requirements.*
>
> > *"Within every industry there are significant leaders and laggards in terms of technological sophistication, software quality and productivity levels. For software, one of the most significant differences is that the leaders know their quality and productivity levels because they measure them...investment in good quality and productivity measurement programs has been demonstrated to be one of the best returns on investment of any known software technology."**

Purpose

This section should describe the objectives of measurement and analysis activities and how these activities support the software process. An overview of all measures, data collection and storage mechanisms, analysis techniques, and reporting and feedback mechanisms should be provided. An example is provided:

> *The purpose of this SMM Plan is to establish, document, and maintain consistent methods for measuring the quality of all software products developed and maintained by the [Project Abbreviation] development team. This plan defines the minimum standardized set for information gathering over the software life cycle. These measures serve as software and system measures and indicators that critical technical characteristics and operational issues have been achieved.*

---

*Capers Jones, The Pragmatics of Software Process Improvement. *Software Process Newsletter,* No. 5, Winter 1996, pp. 1–4.

Scope

This section should provide a brief description of scope to include the identification of all associated projects and a description of measure methodology employed. An example is provided below:

> *The measures to be utilized by [Project Abbreviation] are described in detail in Section 3. These measures deemed to be the most feasible and cost-effective to initially implement are described in Table 1-2 below.*
>
> *The measures employed by [Project Abbreviation] fall into three general categories, as shown in Table 1-2. Management measures deal with contracting, programmatic, and overall management issues. Requirements measures pertain to the specification, translation, and volatility of requirements. Quality measures deal with testing and other software technical characteristics.*
>
> *Additional measures were evaluated for [Project Abbreviation] Program implementation. Those measures reviewed and deemed "not applicable" are noted as NA.*

Table 1-2. Program Measures

Measure	*Producer*	*Objective*	*Measurement*
Management Category			
Manpower	*Program Manager*	*Track actual and estimated man-hours usage*	*Hours used vs. estimated*
Development Progress	*Project Manager*	*Track computer software units (CSU) planned vs. coded*	*Number of coded CSUs*
Cost	*NA*	*Track S/W expenditures*	*NA*
Schedule	*Project Manager*	*Track schedule adherence*	*Milestone/event slippage*
Computer Resource Utilization	*NA*	*Track planned and actual resource use*	*NA*
Software Engineering Environment (This measure is not covered in Section 3)	*[Company Name] EPG*	*Quantify developer S/W engineering environment maturity*	*Computed # of ISO 9001 requirements*
			(continued)

Table 1-2. Continued

Measure	Producer	Objective	Measurement
Requirements Category			
Requirements Traceability	Project Manager	Track requirements to code	% of requirements traced throughout program life cycle
Requirements Stability	Project Manager	Track changes to requirements	# and % of requirements changed/added per project
Quality Category			
Design Stability	Project Manager	Track design changes Assess code quality	Stability index
Complexity	Not implemented due to manpower constraints		Complexity indices
Breadth of Testing	Project Manager	Track testing of requirements	% requirements tested and % reqmnts passed
Depth of Testing	Project Manager	Track testing of code	Degree of code testing
Fault Profiles	Project Manager	Track open vs. closed anomalies	# and types of faults, average open age

The specific products developed and maintained by the [Project Abbreviation] development team may be found in the [Project Abbreviation] Software Configuration Management (SCM) Plan.

Definitions, Acronyms, and Abbreviations

This section should include all definitions and acronyms used during the development of this plan.

References

This section should include a list of all references used during the development of this plan.

Measures Process Management

This section should describe the process for the definition, selection, collection, reporting of measurement and measures. Using a diagram to help describe the process flow is often helpful. The following provides an example of information typical to this section:

> *The [Project Abbreviation] Program Manager, the [Project Abbreviation] Project Manager, and the [Company Name] Engineering Process Group (EPG) will work together to develop, maintain, and implement an effective software measurements and measures program. The results of all measures should be reported at Program Management Reviews (PMRs), Configuration Control Board (CCB) Reviews and Prerelease/ Validation Reviews. Figure 2-1 provides an overview of the [Project Abbreviation] measures collection and reporting process.*

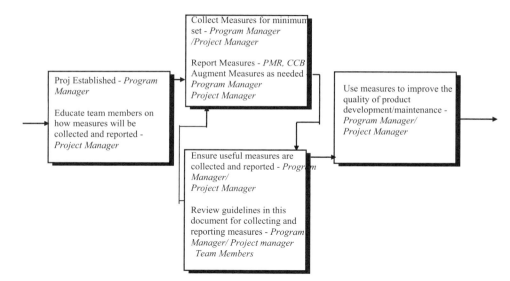

Figure 2-1. [Project Abbrev.] Software Quality Measurement and Measures Process

Responsibilities

This section should describe all responsibilities related to the measurement and measures process. Subsections 2.1.1 and 2.1.2 should provide this information, examples of which follow.

> *Program Manager. The [Project Abbreviation] Program Manager is ultimately responsible for ensuring that Software Measurement and Mea-*

sures (SMM) procedures are developed, implemented, and maintained. In this capacity, the Program Manager will work with the [Company Name] EPG and applicable [Project Abbreviation] CCBs in establishing standards and implementing EPG recommended measures.

The Program Manager is ultimately responsible for ensuring that the [Project Abbreviation] Project Manager and the associated [Project Abbreviation] team are trained in SMM policies and procedures that apply to their function on the team. This will consist of initial training for new members and refresher training, as required, due to process or personnel changes.

Project Manager. The [Project Abbreviation] Project Manager shall provide direct oversight of SMM activities associated with project software development, modification, or maintenance tasks assigned within project functional areas. (Refer to Table 1-2.)

The Project Manager is responsible for ensuring that project SMM are collected and reported to the Program Manager. (Refer to Figure 2-1.)

Life Cycle Reporting

This section should describe all minimum reporting requirements. The following provides an example:

Figure 2-2 below shows the applicability of the minimum measure set over the software life cycle. These measures can provide valuable insight into a program, especially with regard to demonstrated results and readiness for test. The results of all measures should be reported at Program Reviews, Staff (Project) Reviews, and all Validation/Test Reviews. Figure 2-2 also identifies the measures that should be reported at each major decision milestone. Any other measure that indicates the potential for serious problems should also be reported at indicated milestones.

The applicable time periods for data collection and analysis are provided with each measure description. For most measures, data continues to be collected after the system is fielded. The measures collected will be used as program-maturity status indicators. They will be used to portray trends over time, rather than single-point values. When a measures base is established, trends will be compared with data from similar systems.

General Information

This section should provide any additional relevant information not previously provided. The following is an example:

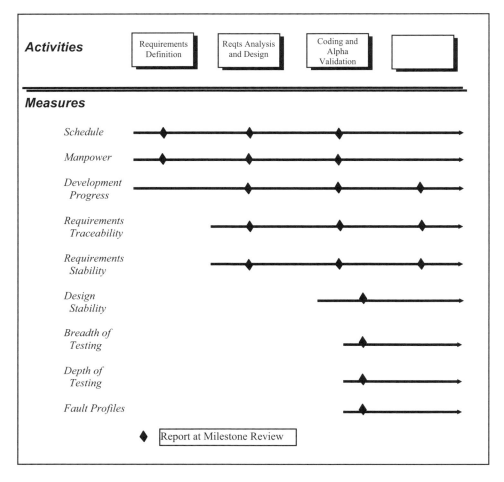

Figure 2-2. Measures and the Project Life Cycle

The graphical displays shown in this plan (see Figure 2-3 on next page) are provided for illustrative purposes. There may be other ways of processing and displaying the data that are more appropriate for a specific [Project Abbreviation] event. Specific project reports will be archived by reporting date by the Program Manager or designee.

Measurement and Measures

This section (the rest of the plan) should provide a detailed description of all measurement collection and measures calculation. This should also include data requirements, the frequency and type of reporting, interpretation recommendations, and examples when possible. It is often helpful to break this down by categories

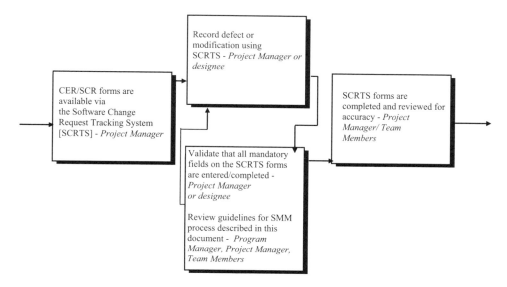

Figure 2-3. Collecting Measures for Each Activity

(e.g., management, requirements, quality). Detailed information should be provided in support of all measures defined within a given category. Example measures are provided in Appendix C, Example Measurements, of this text.

PROBLEM RESOLUTION

The problem resolution process analyzes and resolves the problems (including non-conformances), whatever their nature or source, that are discovered during the execution of development, operation, maintenance, or other processes. Normally, problems found in products under author control are resolved by the author, not in this process. Table 6-14 provides a list of the problem resolution process objectives.

ISO 9001 Goals

ISO 9001 describes requirements in support of the IEEE 12207 problem resolution process. The key requirement is Clause 8.3, Control of Nonconforming Product, to ensure that any nonconforming software product (or erroneous configuration item)

Table 6-14. Validation process objectives

a) Provide a timely, responsive, and documented means to ensure that all discovered problems are analyzed and resolved.

b) Provide a mechanism for recognizing and acting on trends in problems identified.

is identified and controlled to prevent its unintended use or delivery. Records of the nature of the nonconformity, and any subsequent actions (including any concessions), must be maintained. The complimentary requirement is Clause 8.5.2, Corrective Action, to eliminate the cause of the nonconformity and prevent recurrence. The corrective action must be appropriate to effects of the problem.

As errors can occur throughout the project life cycle, the first requirement impacted is Clause 7.1, Planning of Product Realization, to plan and develop processes. The planning must also determine, as appropriate, the verification, validation, monitoring, inspection, and test activities, which provide errors for the problem resolution process. Clause 7.3.1, Design and Development Planning, must plan and control the product design and development. This planning must determine the appropriate review, verification, and validation activities for each stage, which provide errors for the problem resolution process. Clause 7.3.4, Design and Development Review, supports the performance of systematic reviews of design and development in order to identify problems and propose any necessary actions, which provide errors as inputs to the problem resolution process. Clause 7.5.1, Control of Production and Service Provision, must plan and carry out production and service provision under controlled conditions to include, as applicable, implementation of release, delivery, and postdelivery activities, which could include problem resolution, help-desk support, hardware support, and system monitoring to detect failures.

Risk Management Plan

Risk management focuses on the assessment and control (mitigation) of associated project risk. Risk assessment essentially involves the identification of all of the potential dangers, and their severity, that will affect the project and the evaluation of the probability of occurrence and potential loss associated with each item listed. The control of risk requires risk mitigation through the development of techniques and strategies for dealing with the highest-ordered risks. Risk control also involves the evaluation of the effectiveness of the techniques and strategies employed throughout the project lifecycle. Table 6-15 provides a suggested outline for a risk Management plan.

The following provides section-by-section guidance in support of the creation of a risk management plan. ISO 9001 requires that all risks be identified and controlled. Additional information is provided in the document template, *Risk Management Plan.doc,* which is located on the companion CD-ROM. This risk management plan template is designed to facilitate the development of a planned approach to risk management activities. This document template used IEEE Std 1540, IEEE Standard for Software Life Cycle Processes- Risk Management [37], and IEEE Std 12207.0 and 12207.1, Standards for Information Technology—Software Life Cycle Processes [39], as primary reference material that was adapted to support ISO 9001 requirements. Additional information in support of risk-management-associated work products is provided in Appendix C.

Table 6-15. Risk management
plan document outline

Title Page
Revision Page
Table of Contents
1. Introduction
2. Purpose and Scope
3. Definitions and Acronyms
4. References
5. Risk Management Overview
6. Risk Management Policies
7. Organization
8. Responsibilities
9. Orientation and Training
10. Costs and Schedules
11. Process Description
12. Process Evaluation
13. Communication
14. Plan Management

Introduction

This section should provide the reader with a basic description of the project, including associated contract information and any standards used in the creation of the document. A description of the document approval authority and process should be provided.

Purpose

This section should provide a description of the motivation for plan implementation and the expectations for anticipated benefit. Define the perceived benefit due to plan execution.

Scope

This section should describe the specific products and the portion of the software life cycle covered by the risk management plan. An overview of the major sections of the risk management plan should also be provided.

Definitions and Acronyms

This section should provide a complete list of all acronyms and all relevant key terms used in the document that are critical to reader comprehension.

References

List all references used in the preparation of this document in this section.

Risk Management Overview

This section should describe the project's organization structure, its tasks, and its roles and responsibilities. A reference to any associated software project management plan should be provided here.

Risk Management Policies

This section should identify all organizational risk management policies. Refer to IEEE Std 1540 for detailed information supporting risk management requirements.

Organization

This section should depict the organizational structure that assesses and controls any associated project risk. This should include a description of each major element of the organization together with the roles and delegated responsibilities. The amount of organizational freedom and objectivity to evaluate and monitor the quality of the software, and to verify problem resolutions, should be clearly described and documented. In addition, the organization responsible for preparing and maintaining the risk management plan should be identified.

Responsibilities

This section should identify the specific organizational element that is responsible for performing risk management activities. Individuals who typically participate in risk management activities and should be described are program manager, project manager, risk management manager, software team members, quality assurance manager, and configuration management manager.

Orientation and Training

This section should describe the orientation or training activities required in support of risk management activities. If this information is supplied in a related training plan, a reference to this plan may be provided here.

Costs and Schedules

Describe all associated cost and schedule impacts in this section. If this information is supplied in a related software project management plan, a reference to this plan may be provided here.

Process Description

Describe the risk management processes employed by the project in this section. If an organizational risk management process exists, then a reference to these processes may be provided here. If any adaptations are required during the adoption of organizational processes at the project level, then these adaptations must be described in this section. A diagram is often useful when communicating process descriptions.

All techniques and tools used during the risk management process should be described. Risk taxonomies, forms, and databases are often used in support of project-level risk management activities. If these items are used, a description should be provided.

Process Evaluation

Describe how measurement information will be collected in support of the evaluation of the risk management process. It this information is provided in a related software measurement and measurements plan, and then a reference to this plan may be provided here.

Communication

This section should describe how all associated project risk information would be coordinated and communicated among all project stakeholders. It is important to address when risk elevation procedures will take place.

	Consequences				
	Insignificant	Minor	Moderate	Major	Catastrophic
Likelihood	1	2	3	4	5
A (Almost certain)	H	H	E	E	E
B (Likely)	M	H	H	E	E
C (Possible)	L	M	H	E	E
D (Unlikely)	L	L	M	H	E
E (Rare)	L	L	M	H	H

E Extreme Risk – Immediate action; senior management involved
H High Risk – Management responsibility should be specified
M Moderate Risk – Manage by specific monitoring or response
L Low Risk – Manage by routine process

Figure 6-5. Probability/impact risk rating matrix.

Plan Management

The management of the risk management plan, including revisions, approval, and associated configuration management requirements, should be addressed in this section.

Probability/Impact Risk Rating Matrix

IEEE Std 1490, IEEE Guide—Adoption of the Project Management Institute Standard, A guide to the Project Management Body of Knowledge (PMBOK) [34], describes risk management as one of the project management knowledge areas. The PMBOK dedicates an entire chapter to the definition, analysis, and appropriate response to associated project risk. The phrase used by the PMBOK to describe the evaluation of project risk and associated likely outcomes is "risk quantification." Several methods for risk assessment are presented in the PMBOK. These methods range from statistical summaries to expert judgment. The probability/impact risk rating matrix (Figure 6-4) is provided as an example of a way to leverage expert judgment and rate identified project risk.

Chapter 7

12207 Organizational Processes and ISO 9001

IEEE 12207 describes four organizational life cycle processes: the management process, infrastructure process, improvement process, and training process. These processes work at the organizational level for all projects and should be in place prior to performing the five primary life cycle processes of acquisition, supply, development, maintenance, and/or operation, and the eight supporting processes (Figure 7-1). The organization employing and performing a primary and supporting process, needs to

- Manage it at the project level following the management process
- Establish an infrastructure under it following the infrastructure process
- Tailor it for the project following the tailoring process
- Manage it at the organizational level following the improvement process and the training process
- Assure its quality through joint reviews, audits, verification, and validation

Organizational life cycle processes evaluate whether a new, changed, or outsourced process could be supported.

ISO 9001 GOALS

ISO 9001 describes requirements in support of the IEEE 12207 organizational life cycle processes through three ISO 9001 sections: management responsibility, resource management and measurement, and analysis and improvement. These sections provide the foundation for software engineering processes covered in the two previous chapters.

MANAGEMENT

The management process contains the generic activities and tasks managing its respective processes. The manager is responsible for product management, project

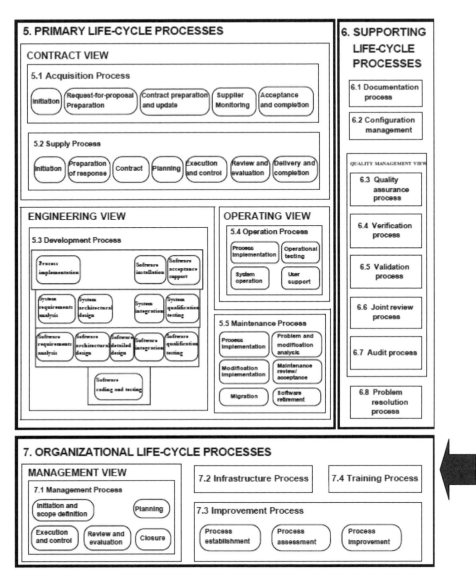

Figure 7-1.

management, and task management of the applicable processes, such as the acquisition, supply, development, operation, maintenance, or supporting process. This process consists of the following activities starting at initiation and scope definition, leading to planning, followed by execution and control, including review and evaluation, and ending with closure. Table 7-1 provides a list of the management process objectives.

Table 7-1. Management process objectives

a) Define the project work scope.
b) Identify, size, estimate, plan, track, and measure the tasks and resources necessary to complete the project.
c) Identify and manage interfaces between elements in the project and with other projects and organizational units.
d) Take corrective action when project targets are not achieved.
e) Establish quality goals, based on the customer's quality requirements, for various checkpoints within the project's software life cycle.
f) Establish product performance goals, based on the customer's requirements, for various checkpoints within the project's software life cycle.
g) Define and use measures that reflect the results of project activities or tasks, at checkpoints within the project's life cycle, to assess whether the technical, quality, and product performance goals have been achieved.
h) Establish criteria, measures, and procedures for identifying software engineering practices and integrate improved practices into the appropriate software life cycle processes and methods.
i) Perform the identified quality activities and confirm their performance.
j) Take corrective action when technical, quality, and product performance goals are not achieved.
k) Determine the scope of risk management to be performed for the project.
l) Identify risks to the project as they develop.
m) Analyze risks and determine the priority in which to apply resources to mitigate those risks.
n) Define, implement, and assess appropriate risk mitigation strategies.
o) Define, apply, and assess risk measures to reflect the change in the risk state and the progress of the mitigation activities.
p) Establish an environment that supports effective interaction between individuals and groups.
q) Take corrective action when expected progress is not achieved.

ISO 9001 Goals

ISO 9001 describes requirements in support of the IEEE 12207 management process. The organization uses the management process to demonstrate its ability to consistently provide software product that meets customer and applicable regulatory requirements, and to enhance customer satisfaction through the effective application of the management system, including processes for continual improvement and the assurance of conformity to customer and applicable regulatory requirements.

From ISO 9000:2000, Quality Management Systems—Fundamentals and Vocabulary, the role of top management is to:

a) Establish and maintain the quality policy and quality objectives of the organization.

b) Promote the quality policy and quality objectives throughout the organization to increase awareness, motivation, and involvement.

c) Ensure focus on customer requirements throughout the organization.

d) Ensure that appropriate processes are implemented to enable requirements of customers and other interested parties to be fulfilled and quality objectives to be achieved.

e) Ensure that an effective and efficient quality management system is established, implemented, and maintained to achieve these quality objectives.

f) Ensure the availability of necessary resources.

g) Review the quality management system periodically.

h) Decide on actions regarding the quality policy and quality objectives.

i) Decide on actions for improvement of the quality management system.

ISO 9001 clauses for the management process are:

Clause 4.1, General Requirements, requires a management system to be established, documented, implemented, and maintained. To implement the system, an organization must determine the needed processes and their sequence and interaction, and determine criteria and methods for process operation and control. To maintain the system, ensure that resources and supporting information are available, and that processes are monitored, measured, and analyzed. Finally, implement actions to achieve planned results and continual process improvement.

Clause 5.1, Management Commitment, requires that senior management provide evidence of its commitment to develop and implement the management system, as well as, continually improve its effectiveness.

Clause 5.2, Customer Focus, requires that senior management ensure that customer requirements are determined and met in order to improve customer satisfaction.

Clause 5.3, Quality Policy, requires that senior management ensure that the quality policy is appropriate, focused, used, communicated, understood, and reviewed for continuing suitability.

Clause 5.4.1, Quality Objectives, requires that senior management ensure that quality objectives, including those needed to meet product requirements, are established at the relevant functions and levels within the organization. Quality objectives must be measurable and consistent with the quality policy.

Clause 5.4.2, Quality Management System Planning, requires that senior management ensure that management system planning meets the general requirements and quality objectives, and maintains the system integrity when changes are planned and implemented.

Clause 5.5, Responsibility, Authority, and Communication, requires that senior management ensure that responsibilities and authorities are defined and communicated within the organization. These include a management representative and internal communication regarding the effectiveness of the system.

Clause 5.6, Management Review, requires that senior management review the management system at planned intervals.

Clause 6.1, Provision of Resources, requires determining and providing the resources necessary to implement and maintain the management system and to continually improve the effectiveness of the system and thus enhance customer satisfaction by meeting customer requirements.

Clause 7.1, Planning of Product Realization, requires planning the necessary project life cycle processes, documents, and resources, followed up by records as evidence that the processes and resulting product meet requirements.

Clause 7.3.1, Design and Development Planning, requires planning and control. The planning specifies the project life cycle processes, the reviews, verification, and validation activities for each process, and the responsibility and authority for each process. The control includes management for effective communication and the clear assignment of responsibility for the different involved groups, while updating plan documents.

Software Requirements Management Plan

IEEE Standard 830, IEEE Recommended Practice for Software Requirements Specification [6], provides detailed guidance in support of the development of a software requirements management plan. This standard provides detail on what is required to effectively manage software requirements and how to document these requirements in a management plan.

IEEE Std 12207.0, Standard for Information Technology—Software Life Cycle Processes [39], describes 17 processes spanning the entire life cycle of a software product or service. Even if an organization's processes were defined using other sources, the standard is useful in characterizing the essential characteristics of these software processes and should be considered prior to the implementation of process improvement activities. Referencing this standard, and reviewing what is required for each of these primary process areas, can provide valuable additional guidance in support of the activities associated with requirements management and analysis. It is important that IEEE 12207 be considered prior to the implementation of process improvement activities associated with requirements management and requirements analysis.

The implementation of IEEE Std 830, Recommended Practice for Software Requirements Specifications, assures the development of a software requirements specification but does not support the requirement of the institutionalization of software requirements management. At a minimum, IEEE Std 830 must be used in conjunction with IEEE Std 1058, Standard for Software Project Management Plans, and IEEE Std 12207.0, Standard for Information Technology—Software Life Cycle Processes, in support of this goal. Incorporating requirements management into the project planning, organizational processes, and management helps to ensure that it is integrated into the life cycle of the project.

In order for the IEEE Std 830, IEEE Recommended Practice for Software Requirements Specifications [6], to be an effective instrument for the implementation of ISO 9001, information regarding the management of the defined requirements needs to be added and stated explicitly. Information needs to be included in a software requirements management plan (SRMP) that will demonstrate that the resources, people, tools, funds, and time have been considered for the management of the requirements that are defined.

This information should relate directly to the software project management plan (SPMP). For example, if one person is required to perform the duties of the requirements manager, this should be identified in the SPMP. This section could reference the SPMP if the information is contained in that document. However, the present IEEE standard for SPMPs does not require the identification of requirements management issues such as resources and funding, reporting procedures, and training.

The recommended SRMP table of contents to support the goals of ISO 9001 more directly is shown in Table 7-2.

Software Requirements Management Plan Document Guidance

The following provides section-by-section guidance in support of the creation of a software requirements management plan. The development of an organizational SRMP can not only help baseline the process of software requirements definition and management, but the SRMP can also provide valuable training and guidance to project performers. This guidance should be used to help define a management process and should reflect the actual processes and procedures of the implementing organization. Additional information is provided in the document template, *Software Requirements Management Plan.doc,* which is located on the companion CD-ROM. Additional information in support of requirements traceability is provided in Appendix C, Work Products.

Purpose

This section should describe the purpose of the software requirements management plan and why effective requirements management is critical to the success of any software development or maintenance effort.

Software Requirements Management activities are concerned with two major areas:

1. Ensuring all software requirements are clearly defined.
2. Preventing the requirements process from becoming burdened by design, verification, or management details that would detract from specifying requirements.

Table 7-2. Software requirements management plan document outline

Title Page
Revision Page
Table of Contents
 1.0 Introduction
 2.0 Purpose
         2.1 Scope
         2.2 Definitions
         2.3 Goals
 3.0 References
         3.1 Key Acronyms
         3.2 Key Terms
         3.3 Key References
 4.0 Management
         4.1 Organization
         4.2 Tasks
         4.3 Responsibilities
                 4.3.1 Management
                 4.3.2 Program Manager
                 4.3.3 Project Lead
                 4.3.4 Team Members
                 4.3.5 Customer
 5.0 Software Requirements Management Overview
         5.1 Software Requirements Modeling Techniques
                 5.1.1 Functional Analysis
                 5.1.2 Object-Oriented Analysis
                 5.1.3 Dynamic Analysis
         5.2 Software Requirements Management Process
                 5.2.1 Requirements Elicitation
                 5.2.2 Requirements Analysis
                 5.2.3 Requirements Specification
         5.3 Characteristics of a Good SRS
                 5.3.1 Correct
                 5.3.2 Nonambiguous
                 5.3.3 Complete
                 5.3.4 Verifiable
                 5.3.5 Consistent
                 5.3.6 Modifiable
                 5.3.7 Traceable
                 5.3.8 Usable during Operation and Maintenance
 6.0 Standards and Practices
 7.0 Software Measurement
 8.0 Verification and Validation
 9.0 Software Configuration Management
10.0 Developing a Software Requirements Specification
Appendix A. Project Software Requirements Specification Template
Appendix B. Requirements Traceability Matrix

Scope

This subsection should address the scope of the plan and its applicability.

Definitions

This subsection should list all relevant defined terminology.

Goals

This subsection should identify all organizational goals associated with software requirements management.

References

A list of key acronyms, definitions, and references should be provided in this section of the document.

Organization

This subsection should describe the individuals responsible for software requirements management. The following is provided as an example:

> *The [PROJECT ABBREVIATION] program manager will ensure completion of Software Requirements Management by coordinating the [PROJECT ABBREVIATION] team development of an SRS. The program Manager will ensure that the training and time is provided to the project leader to have the SRS properly developed. The SRS will be developed to ensure that requirements are properly captured.*

Tasks

This subsection should describe the high-level tasks associated with software requirements management and the specification of requirements, the following is provided as an example:

> *The tasks to be performed in the development of an SRS are:*
>
> 1) *Specifying software requirements as completely and thoroughly as possible;*
> 2) *Defining software product functions;*
> 3) *Defining design constraints;*
> 4) *Creating software Requirements Traceability Matrix;*
> 5) *Defining assumptions and dependencies.*

Responsibilities

This subsection should describe the responsibilities of those individuals at all levels within the organization. It should address the roles of senior management, program, project lead, team members, and the customer.

Software Requirements Management Overview

This section should provide an overview of the software requirements management process. The following is provided as an example:

> *Software requirements management is the set of processes (activities and tasks) and techniques used to define, analyze, and capture the customer's needs for a software-based system. This process starts with the customer's recognition of a problem and continues through the development and evolution of a software product that satisfies the customer's requirements. At that point, the software solution is terminated. For requirements management to work effectively, the requirements must be documented and agreed to by both the customer and [COMPANY NAME]. The formal document that contains this agreement is an SRS. The focus of the requirements management effort is then to create, approve, and maintain an SRS that meets and continues to meet the customer's needs.*

Software Requirements Modeling Techniques

This subsection should provide a description of the software requirements modeling techniques.

Software Requirements Management Process

This subsection should provide a detailed description of the software requirements management process. The following information is provided as an example:

> *The process of creating and approving an SRS can consist of three activities: requirements elicitation, requirements analysis, and requirements specification. Some problems are so clearly stated that only requirements specification is needed; other problems require more refinement and, therefore, also require the elicitation and analysis. This description will cover all three activities, assuming a worst-case scenario.*
>
> *Projects following this plan will assess the state of their requirements and enter the process at the appropriate point. (Include a figure that represents the requirements management process, with all key players identified at appropriate entry points.)*

Requirements Elicitation

Requirements elicitation is the process of extracting information about the product from the customer. Requirements elicitation is highly user interactive. A series of activities may be used to help both the customer and the developer create a clear picture of what the customer requires. These activities are discussed below. They include interviews, scenarios, ambiguity resolution, context analysis, rapid prototyping, and clarification and prioritization. This particular ordering of techniques is most useful for moving from the most obscure request to the most definitive. Other orderings also will work well and should be adapted to fit the project.

Interviews

Interviews are the most natural form of gaining an understanding of the customer's request. However, interviews are often performed with little advance preparation; this is counterproductive. The customer becomes frustrated at being asked unimportant questions; the developer is forced to make assumptions about what is not understood. One way to reduce the failure rate of interviews is to focus the interview using the following strategy:

1. Lose your preconceptions regarding the request. Assuming that you know what must be accomplished prevents you from listening carefully to the customer and may give the customer the perception that the interview is just a formality.
2. Prepare for the interview. Find out what you can about the customer, the people to be interviewed, and the request in general. This gives you a background for understanding the problem. Clarify those areas of uncertainty with the customer; this may help to better formulate requirements.
3. Ask questions that will help you understand the request from a design perspective. These questions lead to clarification of for whom the project is being built and why it is being built. This is similar to the current TV commercial that asks, "How did your broker know you wanted to retire in ten years?" The answer is, "He asked!" These questions might include:

   a) Who is the client? This addresses for whom the system is really being built.
   b) What is a highly successful solution really worth to this client? This addresses the relative value of the solution.
   c) What is the real reason for this request? This might lead to alternative approaches that are more satisfactory to the customer.
   d) Who should be on the development team? This may identify particular skills or knowledge that the customer recognizes as necessary.
   e) What are the customers' trade-offs between time and value? This gives a better perspective on urgency.

4. Ask questions that will help you understand the product better. These might include:

   a) What problem does this product solve?

   b) What problem could this product create? It is a good idea to think about this early. This might also help to bound the problem area.

   c) What environment is this product likely to encounter? Again, this helps to bound the request, and it may help identify some of the restrictions on the solution.

   d) What kind of precision is required or desired in the product?

Scenarios

A scenario is a form of prototyping used to describe a customer request. In a scenario, examples of how a solution might react are examined. Again, this is a very natural process, but one that needs to be prepared in order to work effectively. Scenarios are built by determining how implementation might satisfy the request. Assumptions, problems, and the customer's reactions to the scenarios are used for further evaluation.

Once the interviews and scenarios are progressing, a work statement can be formed. A work statement is a textual description of the customer's interpretation of the request. This may range from very formal to very abstract. The purpose of interviews and scenarios is to clarify the work statement.

Ambiguity Resolution

Ambiguity resolution is an attempt to identify all the possible misinterpretable aspects of a work statement. Examples of ambiguous areas are: terms, the definitions of which are often interpreted differently; acronyms that are not clearly defined; or misleading statements of what is required. Question everything; nothing should be assumed while reading the work statement.

The output of this step is a redefined work statement, possibly with a list of definitions. The list of definitions should be the starting point for a future data dictionary.

Context Analysis

Context analysis is an early modeling of the system to identify key objects as named links. This helps to organize pieces of the request, and shows connectivity to highlight communication.

Clarification and Prioritization

The final step in requirements elicitation is to create a list of the desired functions, attributes, constraints, preferences, and expectations. This moves the work statement toward a more formal software requirements specification.

1. Functions are the "what" of the product. They describe what the product must do or accomplish to satisfy the customer's request. Functions should be classified as *evident, hidden,* and *frill.* Once this classification is complete, the developer has set the stage for what must be demonstrated to the customer (*evident* functions), and what must be developed to satisfy the customer (*hidden* functions). The *frills* are kept on a standby list to be obtained, if possible.

2. Attributes are characteristics of the product. They describe the way the product should satisfy customer needs. Attributes and attribute lists (linear descriptions of how the attribute is recognized) can be classified as *must, want,* and *ignore.* Attributes and attribute lists are attached to the functions described above. To satisfy the customer all *must* attributes have to be met, and as many *want* attributes as possible should be met. *Ignore* attributes are ignored.

3. Constraints are mandatory conditions placed on an attribute. These may come from regulations, policies, or customer experience. (Developers should always attempt to recognize and validate all constraints. Constraints limit the options available and they should be investigated before they limit the system.)

4. Preferences are desirable, but optional, conditions placed on an attribute. Preferences help to satisfy the customer, so they should be achieved whenever possible.

5. Expectations are the hardest items to capture. These are what the customer expects from the product; they are seldom clearly defined. Every attempt should be made to determine what the customer expects the system to do, and these expectations should be met if possible. If the customer's expectations are exorbitant, then these issues should be addressed at the project onset to avoid a late disappointment.

The output of the requirements elicitation process is a well-formed, redefined work statement clearly stating the customer's request. This work statement provides the basis for the requirements analysis.

Requirements Analysis

Requirements analysis involves taking the work statement and creating a model of the customer's request. The types of request should determine which technique to select.

Requirements Specification

The specification, called the SRS, is the formal definition of the statement to the customer's problem. The requirements specification combines the knowledge gained during requirements elicitation (attributes, constraints, and preferences) with the model produced during requirements analysis.

Characteristics of a Good SRS

This section should describe the characteristics of a good SRS. The following are offered as criteria:

Correct. "An SRS is correct if and only if every requirement stated therein represents something required of the system to be built." [69]

Nonambiguous. "An SRS is nonambiguous if and only if every requirement stated therein has only one interpretation." [69]

Complete. "An SRS is complete if it possesses the following four qualities:

1. Everything that the software is supposed to do is included in the SRS.
2. All definitions of software responses to all realizable classes of input data, in all realizable classes of situations.
3. All pages are numbered; all figures and tables are numbered, named, and referenced; all terms and units of measure are provided; and all referenced material and sections are present.
4. No sections are marked "To Be Determined" (TBD). [69]

Verifiable. An SRS is verifiable only if every requirement stated therein is verifiable. A requirement is verifiable if there exists some finite cost-effective process with which a person, or machine, can check that the actual as-built software product meets the requirement.

Consistent. An SRS is consistent if and only if no subset of individual requirements stated therein conflict. This may be manifested in a number of ways:

1. Conflicting terms
2. Conflicting characteristics
3. Temporal inconsistency*

Modifiable. "An SRS is modifiable if its structure and style are such that any necessary changes to the requirements can be made easily, completely, and consistently." [69]

Traceable. "An SRS is traceable if the origin of each of its requirements is clear and if it facilitates the referencing of each requirement in future development or enhancement documentation." [69]

Usable during Operation and Maintenance. This is a combination of being modifiable as described above and being understandable to users of the SRS. The SRS should last as long as the system built from it, and always accurately reflect the state of the system. SRS requirements that do not meet the above tests should be considered a risk or a reportable issue for the project manager's open issues list.

---

*Temporal inconsistency is when two or more parts of the SRS require the product to obey contradictory timing constraints.

Standards and Practices

This section should provide references to applicable corporate policies and procedures and external standards if appropriate. The following is provided as an example:

> *[PROJECT ABBREVIATION] Projects will document and maintain the current status of all customer approved SRSs. At this time, the standard for software requirements management is limited to the need to establish and follow the format of Appendix A to create specifications for each new project. This format will be followed unless the program is required, by contract, to use another format. As [Company Name]'s maturity within the Capability Maturity Model increases, more detailed standards and practices will evolve. These standards will be based on measurements to ensure consistent capability to complete the requirements.*
>
> *In developing the software requirements specification, the program manager will incorporate all existing [company name] standards and policies for software, including, but not limited to software configuration management and software quality assurance.*

Software Measurement

This section should describe all measurement activities associated with an organizations' management of software requirements. The following is provided as an example:

> *Software measurements will be applied to the process of software requirements management on a continual basis in order to ensure that customer requirements are met and to improve the SRM process. Measurement from a [PROJECT ABBREVIATION] perspective means measuring progress toward the [PROJECT ABBREVIATION] goal of accurately identifying project software requirements. (Reference project level software measurement plan if appropriate.)*
>
> *In order to improve, the results of projects must be gathered, and include deviations from the SRSs. Recording and learning from the lessons of early projects will improve the software requirements management process and serve to move [PROJECT ABBREVIATION] toward the goal of effective requirements management. Both corporate and project configuration management will help review SRSs, and identify deviations.*

Verification and Validation

This section should describe the minimum acceptable verification and validation of software requirements. Verification and validation are two important aspects of any project software life cycle. They provide the checks and balances to the development process that ensure that the system is being built (validation). According to IEEE/EIA 12207.0, verification is "confirmation by examination and provision of objective evidence that specified requirements have been fullfilled," and validation is "confirmation by examination and provision of objective evidence that the particular requirements for a specific intended use are fullfilled." [2] In other words, verification is "Are we building the right product?" and validation is "Are we building the product right?"

Verification and validation differ in scope and purpose. Verification is concerned with comparing the results of each activity of the software life cycle with the specification from the previous activity. The scope of verification, therefore, is limited to two activities (current and prior). For this reason, the context of verification changes depends upon the pair of activities under review.

Validation takes a birds-eye view of requirements and the implementation of those requirements. Each functional and nonfunctional requirement listed in the SRS will be testable by one of four techniques. The SRS will reference a requirements traceability matrix (RTM), that lists all requirements and the method used to validate that requirement. Four possible methods of validation are:

1. **Demonstration.** The end product is executed for the customer to demonstrate that the desired properties have been met.
2. **Test.** The end product is executed through predefined, scripted tests for the customer, to show functionality and demonstrate correctness.
3. **Analysis.** Results of the product execution are examined to ensure program correctness.
4. **Inspection.** The product itself is inspected to demonstrate that it is correct.

Software Configuration Management

This section should provide a cross-reference to the supporting software configuration management plan and associated activities. The following is provided as an example:

> The software project management plan will present a schedule for developing the SRS. The schedule will indicate the expected time for baselining the completed SRS. The baselined SRS will be placed under configuration management defined in the program Configuration Management Process.

Developing a Software Requirements Specification

This section should provide detailed guidance in support of the development of a SRS. The following is provided as guidance:

> Each program manager should ensure that all new projects develop an SRS and submit it through the Corporate SQA Manager to [COMPANY

NAME] Management for approval. Within each of the sections listed in Appendix A, a general explanation of the section contents is given. Appendix A should be considered a boilerplate for developing your SRS. If a section does not pertain to your project, the following should appear below the section heading "This section is not applicable to this plan," together with appropriate reasons for the exclusion. Additional sections and appendices may be added as needed. Some of the required material may appear in other project documents. If so, then references to these documents should be made in the body of the software requirements specification. In any case, the contents of each section of the plan should be specified either directly or by reference to another document.

Appendix A. Project Software Requirements Specification

This appendix should contain a document template for project use.

Appendix B. Template for Requirements Traceability Matrix

This appendix should contain a traceability matrix template for project use.

Software Project Management Plan

IEEE Std 1058, IEEE Standard for Software Project Management Plans [15], and IEEE/EIA 12207.0, IEEE Standard for Life Cycle Processes [39], are effective instruments of the ISO 9001 requirements in support of software project planning. However, information regarding the measurements required in support of software project planning activities needs to be added and stated explicitly. It is also important to note that though the requirements are addressed by the major headings of IEEE 1058, the details required to support ISO 9001 can be lost if each section is not carefully addressed while bearing the specific ISO 9001 project planning commitments requirements in mind. The modification of the recommended software project management plan (SPMP) table of contents to support the goals of ISO 9001 more directly is shown in Table 7-3.

Software Project Management Plan Document Guidance

The following provides section-by-section guidance in support of the creation of a SPMP. The SPMP should be considered to be a living document. The SPMP should change, in particular, any associated schedules and to reflect any required change during the life cycle of a project. This guidance should be used to help define a management process and should reflect the actual processes and procedures of the implementing organization. Additional information is provided in the document template, *Software Project Management Plan.doc,* that is located on the companion

Table 7-3. Software project management
plan document outline

Title Page
Revision Page
Table of Contents
1. Introduction
    1.1 Project Overview
    1.2 Project Deliverables
    1.3 Document Overview
    1.4 Acronyms and Definitions
2. References
3. Project Organization
    3.1 Organizational Policies
    3.2 Process Model
    3.3 Organizational Structure
    3.4 Organizational Boundaries and Interfaces
    3.5 Project Responsibilities
4. Managerial Process
    4.1 Management Objectives and Priorities
    4.2 Assumptions, Dependencies, and Constraints
    4.3 Risk Management
    4.4 Monitoring and Controlling Mechanisms
    4.5 Staffing Plan
5. Technical Process
    5.1 Tools, Techniques, and Methods
    5.2 Software Documentation
    5.3 Project Support Functions
6. Work Packages
    6.1 Work Packages
    6.2 Dependencies
    6.3 Resource Requirements
    6.4 Budget and Resource Allocation
    6.5 Schedule
7. Additional Components

CD-ROM. Additional information is provided in Appendix C, Work Products, that describes the work breakdown structure, workflow diagram, and stakeholder involvement matrix work products.

Project Overview

This subsection should briefly state the purpose, scope, and objectives of the system and the software to which this document applies. It should describe the general nature of the system and software; summarize the history of system development, operation, and maintenance; identify the project sponsor, acquirer, user, developer, and support agencies; and identify current and planned operating sites. The project overview should also describe the relationship of this project to other projects, as appropriate, addressing any assumptions and constraints. This subsection should also provide a brief schedule and budget summary. This overview should not be construed as an official statement of product requirements. Reference to the official statement of product requirements should be provided in this subsection of the SPMP.

Project Deliverables

This subsection of the SPMP should list all of the items to be delivered to the customer, the delivery dates, delivery locations, and quantities required to satisfy the terms of the project agreement. This list of project deliverables should not be construed as an official statement of project requirements.

Document Overview

This subsection should summarize the purpose and contents of this document and describe any security or privacy considerations that should be considered to be associated with its use. This subsection of the SPMP should also specify the plans for producing both scheduled and unscheduled updates to the SPMP. Methods of disseminating the updates should be specified. This subsection should also specify the mechanisms used to place the initial version of the SPMP under change control and to control subsequent changes to the SPMP.

Acronyms and Definitions

This subsection should identify acronyms and definitions used within the project SPMP. The project SPMP should only list acronyms and definitions used within the SPMP.

References

This section should identify the specific references used within the project SPMP. The project SPMP should only contain references used within the SPMP.

Organizational Policies

This subsection of the SPMP should identify all organizational policies relative to the software project.

Process Model

This subsection of the SPMP should specify the (life cycle) software development process model for the project, describe the project organizational structure, identify organizational boundaries and interfaces, and define individual or stakeholder responsibilities for the various software development elements.

Organizational Structure

This subsection should describe the makeup of the team to be used for the project. All project roles and stakeholders should be identified as well as a description of the internal management structure of the project. Diagrams may be used to depict the lines of authority, responsibility, and communication within the project.

Organizational Boundaries and Interfaces

This subsection should describe the limits of the project, including any interfaces with other projects or programs, the application of the program's SCM and SQA (including any divergence from those plans), and the interface with the project's customer. This section should describe the administrative and managerial boundaries between the project and each of the following entities: the parent organization, the customer organization, subcontracted organizations, or any other organizational entities that interact with the project. In addition, the administrative and managerial interfaces of the project support functions, such as configuration management, quality assurance, and verification, should be specified in this subsection.

Project Responsibilities

This subsection should describe the project's approach through a description of the tasks required to complete the project (e.g., requirements → design → implementation → test) and any efforts (update documentation, etc.) required to successfully complete the project. It should state the nature of each major project function and activity, and identify the individuals, or stakeholders, who are responsible for those functions and activities.

Managerial Process

This section should specify management objectives and priorities; project assumptions, dependencies, and constraints; risk management techniques; monitoring and controlling mechanisms to be used; and the staffing plan.

Management Objectives and Priorities

This subsection should describe the philosophy, goals, and priorities for management activities during the project. Topics to be specified may include, but are not limited to, the frequency and mechanisms of reporting to be used; the relative priorities among requirements, schedule, and budget for this project; risk management procedures to be followed; and a statement of intent to acquire, modify, or use existing software.

Assumptions, Dependencies, and Constraints

This subsection should state the assumptions on which the project is based, the external events the project is dependent upon, and the constraints under which the project is to be conducted.

Risk Management

This subsection should identify the risks for the project. Completed risk management forms should be maintained and tracked by the project leader with associated project information. These forms should be reviewed at weekly staff meetings. Risk factors that should be considered include contractual risks, technological risks, risks due to size and complexity of the project, risks in personnel acquisition and retention, and risks in achieving customer acceptance of the product.

Monitoring and Controlling Mechanisms

This subsection should define the reporting mechanisms, report formats, information flows, review and audit mechanisms, and other tools and techniques to be used in monitoring and controlling adherence to the SPMP. A typical set of software reviews is listed in Appendix C. Project monitoring should occur at the level of work packages. The relationship of monitoring and controlling mechanisms to the project support functions should be delineated in this subsection. This subsection should also describe the approach to be followed for providing the acquirer or its authorized representative access to developer and subcontractor facilities for review of software products and activities.

Staffing Plan

This subsection should specify the numbers and types of personnel required to conduct the project. Required skill levels, start times, duration of need, and methods for obtaining, training, retaining, and phasing out of personnel should be specified.

Technical Process

This section should specify the technical methods, tools, and techniques to be used on the project. In addition, the plan for software documentation should be specified, and plans for project support functions such as quality assurance, configuration management, and verification and validation may be specified.

Tools, Techniques, and Methods

This subsection of the SPMP should specify the computing system(s), development methodology(s), team structures(s), programming language(s), and other notations, tools, techniques, and methods to be used to specify, design, build, test, integrate, document, deliver, and modify or maintain or both (as appropriate) the project deliverables.

This subsection should also describe any tools (compilers, CASE tools, and project management tools), any techniques (review, walk-through, inspection, or prototyping), and the methods (object-oriented design or rapid prototyping) to be used during the project.

Software Documentation

This subsection should contain, either directly or by reference, the documentation plan for the software project. The documentation plan should specify the documentation requirements and the milestones, baselines, reviews, and sign-offs for software documentation. The documentation plan may also contain a style guide, naming conventions, and documentation formats. The documentation plan should provide a summary of the schedule and resource requirements for the documentation effort. IEEE Std for Software Test Documentation (IEEE Std 829) [5] provides a standard for software test documentation.

Project Support Functions

This subsection should contain, either directly or by reference, plans for the supporting functions for the software project. These functions may include, but are not limited to, configuration management, software quality assurance, and verification and validation.

Work Packages

This section of the SPMP should specify the work packages, identify the dependency relationships among them, state the resource requirements, provide the allocation of budget and resources to work packages, and establish a project schedule.

The work packages for the activities and tasks that must be completed in order to satisfy the project agreement must be described in this section. Each work package should be uniquely identified; identification may be based on a numbering scheme and descriptive titles. A diagram depicting the breakdown of activities into subactivities and tasks may be used to depict hierarchical relationships among work packages.

Dependencies

This subsection should specify the ordering relations among work packages to account for interdependencies among them and dependencies on external events. Techniques such as dependency lists, activity networks, and the critical path may be used to depict dependencies.

Resource Requirements

This subsection should provide, as a function of time, estimates of the total resources required to complete the project. Numbers and types of personnel, computer time, support software, computer hardware, office and laboratory facilities, travel, and maintenance requirements for the project resources are typical resources that should be specified.

Budget and Resource Allocation

This subsection should specify the allocation of budget and resources to the various project functions, activities, and tasks. Defined resources should be tracked.

Schedule

This subsection should be used to capture the project's schedule, including all milestones and critical paths. Options include Gantt charts (*Milestones Etc.*™ or *Microsoft Project*™), Pert charts, or simple time lines.

Additional Components

This section should address additional items of importance on any particular project. This may include subcontractor management plans, security plans, independent verification and validation plans, training plans, hardware procurement plans, facilities plans, installation plans, data conversion plans, system transition plans, or product maintenance plans.

Stakeholder Involvement

ISO 9001 requires that all stakeholder involvement be monitored against the project plan. IEEE Std 1058 addresses the identification of all key participants (project stakeholders) associated with a development effort. This standard requires the definition of the organizational structure, all customer relationships, roles, and responsibilities, resources, and how involvement is monitored. ISO 9001 has placed special emphasis on stakeholder involvement in the software development process and as such should be specifically addressed in the software project management plan. Refer to Appendix C for an example of a stakeholder involvement matrix.

Work Breakdown Structure (WBS)

A work breakdown structure (WBS) defines and breaks down all work associated with a project into manageable parts. It describes all activities that have to occur to accomplish the project. The WBS can serve as the foundation for the integration of project component schedules, budget, and resource requirements. IEEE Std 1490,

IEEE Guide—Adoption of the PMI Standard, A Guide to the Project Management Body of Knowledge [34], recommends the use of the structure shown in Figure 7-2. Additional information in support of associated IPM work products is provided in Appendix C, Work Products.

A work breakdown structure (WBS) defines and breaks down the work associated with a project into manageable parts. It describes all activities that have to occur to accomplish the project. The WBS serves as the foundation for the development of project schedules, budget, and resource requirements.

A WBS may be structured by project activities or components, functional areas or types of work, or types of resources, and is organized by its smallest component—a work package. A work package is defined as a deliverable or product at the lowest level of the WBS. Work packages may also be further subdivided into activities or tasks. IEEE Std 1490, IEEE Guide—Adoption of the PMI Standard, A Guide to the Project Management Body of Knowledge, recommends the use of nouns to represent the "things" in a WBS. Figure 7-3 provides an example of a sample WBS organized by activity.

Work Breakdown Structure (WBS) for Postdevelopment Stage

The example provided in Table 7-4 provides suggested content for a WBS in support of the postdevelopment stage of a software development life cycle. This WBS content is illustrative only and should be customized to meet specific project requirements.

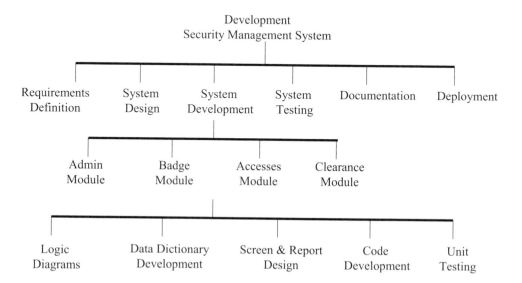

Figure 7-2. Example work breakdown structure.

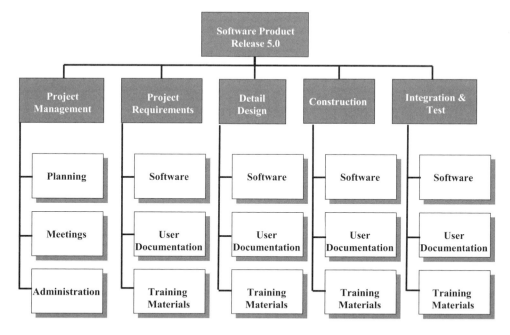

Figure 7-3. WBS organized by activity [34].

INFRASTRUCTURE

The infrastructure process is a process used to establish and maintain the infrastructure needed for any other process. The infrastructure may include hardware, software, tools, techniques, standards, and facilities for development, operation, or maintenance. Table 7-5 describes the infrastructure process objectives.

The infrastructure process objectives include:

a) Establishing and maintaining a well-defined software engineering environment, consistent with, and supportive of, the set of standard processes and organizational methods and techniques.

b) Tailoring the software engineering environment to the needs of the project and the project team.

c) Developing a software engineering environment that supports project team members regardless of the performance location of process activities.

d) Implementing a defined and deployed strategy for reuse.

ISO 9001 Goals

ISO 9001 describes requirements in support of the IEEE 12207 infrastructure process. The single ISO 9001 requirement, Clause 6.3, Infrastructure, requires

Table 7-4. WBS content for postdevelopment activity

Product installation
  Distribute product (software)
    Package and distribute product
    Distribute installation information
    Conduct integration test for product
    Conduct regression test for product
    Conduct user acceptance test for software
    Conduct reviews for product
    Perform configuration control for product
    Implement documentation for product
  Install product (software)
    Install product (packaged software) and database data
    Install any related hardware for the product
    Document installation problems
  Accept product (software) in operations environment
    Compare installed product (software) to acceptance criteria
    Conduct reviews for installed product (software)
    Perform configuration control for installed product (software)
Operations and support
    Utilize installed software system
    Monitor performance
    Identify anomalies
    Produce operations log
    Conduct reviews for operations logs
    Perform configuration control for operations logs
Provide technical assistance and consulting
  Provide response to technical questions or problems
  Log problems
Maintain support request logs
  Record support requests
  Record anomalies
  Conduct reviews for support request logs
Maintenance
  Identify product (software) improvements needs
    Identify product improvements
    Develop corrective/perfective strategies
    Produce product (software) improvement recommendations
  Implement problem reporting method
    Analyze reported problems
    Produce report log
    Produce enhancement problem reported information
    Produce corrective problem reported information
    Perform configuration control for reported information
  Reapply software life cycle methodology

Table 7-5. Infrastructure process objectives

a) Establishing and maintaining a well-defined software engineering environment, consistent with, and supportive of, the set of standard processes and organizational methods and techniques.
b) Tailoring the software engineering environment to the needs of the project and the project team.
c) Developing a software engineering environment that supports project team members regardless of the performance location of process activities.
d) Implementing a defined and deployed strategy for reuse.

the organization to determine, provide, and maintain the necessary infrastructure to achieve product conformity. Infrastructure includes work space, process equipment, and supporting services. The ISO 9001 definition of infrastructure is: the organization or system of facilities, equipment and services needed for the operation of an organization. [47] The main software engineering investments for this requirement is the physical work space and software engineering tools and services.

Organization's Set of Standard Processes

The information provided here is based upon IEEE Std1074, IEEE Standard for Developing Software Life Cycle Processes [19]; and IEEE/EIA Std 12207.0, Industry Implementation of International Standard ISO/IEC 12207 (ISO/IEC 12207), Standard for Information Technology—Software Life Cycle Processes [39]. Table 7-6 describes the three categories of life cycle processes that are found in IEEE 12207.0.

This three-dimensional view of standard processes is supported by Figure 7-4. This figure is from IEEE 12207 and also provides an illustration of representative viewpoints.

IMPROVEMENT

The improvement process is a process for establishing, assessing, measuring, controlling, and improving a software life cycle process. The improvement process uses software life cycle data as it provides a history of what happened during development and maintenance. Table 7-7 describes the improvement process objectives.

ISO 9001 Goals

ISO 9001 describes requirements in support of the IEEE 12207 improvement process. As continual improvement is a principle of ISO 9001, there are eight supporting clauses:

Clause 4.1, General Requirements, requires that the management system must be established, documented, implemented, and maintained. Its effectiveness must be continually improved. All processes must be identified, monitored, measured, and analyzed in order to implement actions for continual process improvement.

Clause 5.6.1, Management Review—General, requires that senior management review the system at planned intervals to assess possible opportunities for improvement.

Clause 8.1, Measurement, Analysis, and Improvement—General, requires the planning and implementation of the monitoring, measurement, analysis, and improvement processes needed to demonstrate product conformity to requirements and process conformity. It also requires that the effectiveness of each be continually improved.

Clause 8.2.2, Internal Audit, requires the organization to conduct periodic internal audits to determine if the management system conforms to planned arrangements and is effectively implemented and maintained. Management must ensure that actions are taken without undue delay to eliminate detected nonconformities and their causes. Follow-up actions must verify implementation of the action and report the results.

Clause 8.2.3, Monitoring and Measurement of Processes, requires suitable methods for monitoring and measurement of the management system processes and for confirmation of each process to continally satisfy its intended purpose.

Clause 8.4, Analysis of Data, requires the organization to determine, collect, and analyze appropriate data to evaluate where continual improvement of the effectiveness of the management system can be made.

Table 7-6. IEEE 12207 life cycle processes [39]

Primary life cycle processes	Acquisition process
	Supply process
	Development process
	Operation process
	Maintenance process
Supporting life cycle processes	Documentation process
	Configuration management process
	Quality assurance process
	Verification process
	Validation process
	Joint review process
	Audit process
	Problem resolution process
Organizational life cycle processes	Management process
	Infrastructure process
	Improvement process
	Training process

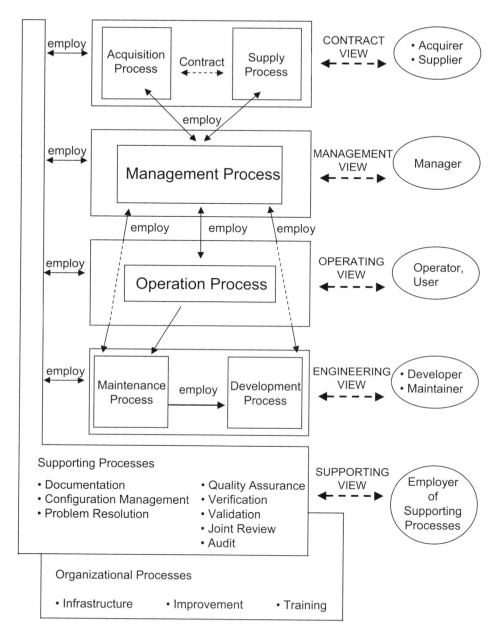

Figure 7-4. Life cycle processes and views [39].

Table 7-7. Improvement process objectives

a) Establish a well-defined and maintained standard set of processes, along with a description of the applicability of each process;
b) Identify the detailed tasks, activities, and associated work products for each standard process, together with expected criteria;
c) Establish a deployed specific process for each project, tailored from the standard process in accordance with the needs of the project;
d) Establish and maintain information and data related to the use of the standard process for specific projects;
e) Understand the relative strengths and weaknesses of the organization's standard software processes;
f) Make changes to standard and defined processes in a controlled way;
g) Implement planned and monitored software process improvement activities in a coordinated manner across the organization.

Clause 8.5.1, Continual Improvement, requires the organization to continually improve the effectiveness of the management system through its policy, objectives, audit results, analysis of data, preventive action, and management review.

Clause 8.5.3, Preventive Action, requires the organization to determine the actions needed to eliminate the causes of potential nonconformities in order to prevent their occurrence.

Engineering Process Group Charter

The following is provided as an example of the type of information typically provided in an engineering process group (EPG) charter document. The information provided here is for illustrative purposes; the EPG charter of any organization should reflect the unique requirements of the developing organization.

Purpose

This section should describe the rationale and scope of the EPG charter. An example is shown:

> *This charter formally empowers the Engineering Process Group (EPG) to manage the released content of corporate software engineering processes at [Company Name].*

Vision/Mission Statement

This section should describe the mission of the EPG, including major areas of responsibility. The text shown provides an example:

The EPG provides management control of [Company Name] engineering processes. It:

a) *Approves release into general use of engineering process assets*
b) *Manages proposed improvements to the engineering process assets as necessary*
c) *Provides tactical guidance on asset creation and maintenance priorities*
d) *Adjudicates controversial or unprecedented process tailoring decisions, as required*

Authority and Responsibility

This section should provide a description of EPG authority and responsibility. An example is provided:

The EPG has authority to define all systems and software engineering process assets in each released set comprising the [Company Name] system and software engineering process. The EPG is responsible for the management and implementation of that asset set.

Membership and Organization

A description of all voting and nonvoting members should be provided. The terms that each member is responsible for serving on the EPG should also be described. It is important to address what may be considered a quorum to convene meetings. An example of addressing the quorum is provided:

Quorum for the meeting shall consist of the Chair, SQA representative, and 60% of the voting membership.

Activities and Function

All activities should be described in this part of the charter document. These activities and functions may include a description of process development, procedures for fast tracking processes, typical review and approval timeline, and supporting procedures.

Process Action Plan (PAP)

Process action plans usually result from appraisals and document how specific improvements targeting the weaknesses uncovered by an appraisal will be implemented. In cases in which it is determined that the improvement described in the process action plan should be tested on a small group before deploying it across the organization, a pilot plan is generated. Finally, when the improvement is to be deployed, a deployment plan is used. This plan describes when and how the improvement will be deployed across the organization.

The following is provided as an example of the type of information typically provided in a process action plan (PAP) document. The information provided here is for illustrative purposes; any specific PAP of an organization should reflect the unique requirements of the developing organization.

Identification

This section should provide summary information that may be used when referring to the document. This should include the author, the date, and any unique tracking number.

Instructions

This section should provide directions for using the PAP, including submission procedures.

Issue

This section should provide a description of the process area being defined or improved. If this is an improvement suggestion, identify the process being used in the context of the proposal being submitted. If a specific tool, procedure, or other EPG asset is being addressed, describe it here as well.

Process Action Team

Identify the personnel to implement the actions.

Process Improvement

This section should provide the strategies, approaches, and actions to address the identified process improvements.

Improvement

This section should describe the process or technological improvement being proposed to support resolution of the issue described above.

Tailoring Guidelines

Processes should be tailored for a project, as no two projects are the same. Tailoring is the adaptation of an organization's set of standard processes for use on a specific project by using tailoring guidelines defined by the organization. The purpose of tailoring the organizational process asset for a specific project is to ensure that the appropriate amount of effort is devoted to the appropriate activities to reduce project risk to an acceptable level while at the same time making most cost-effective use of engineering talent.

IEEE 12207.0 Annex A, Tailoring Process, has four major steps:

1. Identify project environment—strategy, activity, and requirements
2. Solicit inputs—from users, support team, and potential bidders
3. Select processes, activities, documentation, and responsibilities
4. Document tailoring decisions and rationale

Elaborations of these general steps in tailoring the organizational process asset for a project are:

1. Characterize the project environment:
   - Size of the development team
   - Strategy
   - Critical factors
   - Project/customer requirements
   - Life cycle strategy: waterfall, evolutionary, spiral, and so on
   - Life cycle activity: prototyping and maintenance
   - Software characteristics: COTS, reuse, and embedded firmware
   - Project policies, languages, hardware reserve, and culture
   - Acquisition strategy: contract type and contractor involvement
2. Determine the cost targets and risk tolerance of the project.
3. Identify the organization process asset for which tailoring is being considered.
4. For each committed deliverable asset, use best practices and professional judgment to identify the *form* the asset should take and the *level of detail* necessary to achieve the purpose of each organization process asset.
5. Assess whether any organization process asset, or their forms, or their level of detail are unaffordable given the project goals, cost targets, and level of tolerable risk.
6. Identify the level of detail needed for each process activity.
7. Document the tailoring planned to the organization process asset, and obtain approval. Typically, it is unwise to tailor the asset intent or objectives. What normally is tailored are: number of phases/activities, roles, responsibilities, document formats, and formality/frequency of reports or reviews.
8. Document the planned processes, products, and reviews. Describe the completion criteria for each process.

9. After project completion, provide example implementations, lessons learned, and improvements back to the organization to continually improve the organizational process assets.

TRAINING

The training process provides and maintains trained personnel. All software engineering processes are largely dependent upon knowledgeable and skilled personnel. It is, therefore, imperative that training be planned and implemented early so that trained personnel are available as the software product is acquired, supplied, developed, operated, or maintained. The training process objectives are described in Table 7-8.

ISO 9001 Goals

ISO 9001 describes requirements in support of the IEEE 12207 training process. The singular ISO 9001 requirement, Clause 6.2, Human Resources, requires that people performing work affecting product quality must be competent based on the appropriate education, training, skills, and experience. The organization must determine the competency needs for personnel, provide training, evaluate the effectiveness of the actions taken, inform employees of the relevance and importance of their activities, ensure that they appreciate their contribution to achieving quality objectives, and maintain education, training, skill, and experience records.

Training Plan

This document template uses IEEE Std-12207.0, Standard for Information Technology Software Lifecycle Processes; IEEE Std- 12207.1, Standard for Information

Table 7-8. Training process objectives

a) Identify the roles and skills required for the operations of the organization and the project;
b) Establish formal procedures by which talent is recruited, selected, and transitioned into assignments in the organization.
c) Design and conduct training to ensure that all individuals have the skills required to perform their assignments.
d) Identify and recruit or train, as appropriate, individuals with the required skills and competencies to perform the organizational and project roles.
e) Establish a workforce with the skills to share information and coordinate their activities efficiently.
f) Define objective criteria against which unit and individual training performance can be measured, to provide performance feedback and enhance performance continuously.

Technology Software Lifecycle Processes—Life Cycle Data; IEEE Std-12207.2, Standard for Information Technology Software Lifecycle Processes—Implementation Considerations [39]; and U.S. DOE Training Plan Template* as primary reference material. Table 7-9 provides a suggested training plan document outline. Additional information is provided in Appendix C in support of this ISO requirement, in the form of an example training request form and training log.

The following provides section-by-section guidance in support of the creation of a training plan. Ensuring that all individuals are properly trained is integral to the success of any software development effort. Additional information is provided in the document template, *Training Plan.doc,* which is located on the companion CD-ROM. This training plan template is designed to facilitate the development of a planned approach to organizational or project-related training.

Introduction

This section should describe the purpose of the training plan and the organization of the document. As described by IEEE Std 12207.2, Standard for Information Technology Software—Lifecycle Processes—Implementation Considerations:

> The Training Process is a process for providing and maintaining trained personnel. The acquisition, supply, development, operation, or maintenance of software products is largely dependent upon knowledgeable and skilled personnel. For example: developer personnel should have essential training in software management and software engineering. It is, therefore, imperative that personnel training be planned and implemented early so that trained personnel are available as the software product is acquired, supplied, developed, operated, or maintained.

Scope

Describe the scope and purpose of the training, such as initial training for system users, refresher training for the system maintenance staff, training for system administrators, and so on.

Objectives

Describe the objectives or anticipated benefits from the training. It is best to express all objectives as actions that the users will be expected to perform once they have been trained.

Background

Provide a general description of the project, and an overview of the training requirements.

*Department of Energy, Training Plan Template, http://cio.doe.gov/ITReform/sqse/download/traintem.doc.

Table 7-9. Training plan document outline

Title Page
Revision Page
Table of Contents
1.0 Introduction
    1.1 Scope
    1.2 Objectives
    1.3 Background
    1.4 References
    1.5 Definitions and Acronyms
2.0 Training Requirements
    2.1 Roles and Responsibilities
    2.2 Training Evaluation
3.0 Training Strategy
    3.1 Training Sources
    3.2 Dependencies/Constraints/Limitations
4.0 Training Resources
    4.1 Vendor Selection
    4.2 Course Development
5.0 Training Environment
6.0 Training Materials

References

Identify sources of information used to develop this document, including reference to the existing organizational training policy. Reference to the organizational training plan, if the plan being developed is tactical, should be provided as well. Any existing engineering process group (EPG) charter should be referenced to provide insight into process oversight activities, if needed.

Definitions and Acronyms

Provide a description of all definitions and acronyms used within this document that may not be commonly understood, or which are unique in their application within this document.

Training Requirements

Describe the general work environment (including equipment), and the skills for which training is required (management, business, technology, etc.). The training audience should also be identified (category of user: upper management, system administrator, administrative assistant, etc.). It may also identify individuals or positions needing specific training. Include the time frame in which training must be accomplished. Identify whether the training requirements are common or cross-project requirements, or if the requirements are in support of a unique project requirement.

Roles and Responsibilities

Identify the roles and responsibilities of the training staff. Identify the individuals responsible for the management of the training development and implementation. It may also include the identification of other groups that may serve as consultants, such as members of the development team, experienced users, and so on. The individuals responsible for keeping training records and for updating any associated organizational training repository should be identified. If this information is provided in the associated project plan, a reference to this plan may be provided here.

Training Evaluation

The effectiveness of training must continually be evaluated. Describe how training evaluation will be performed. Evaluation tools, surveys, forms, and so on, should be included. Also describe the revision process with regard to the modification of the course and course materials resulting from the evaluations.

Training Strategy

Describe the type of training (e.g., classroom, CBT, etc.) and the training schedule (duration, sites, and dates). Some factors may include adequacy of training facilities; accommodations; need to install system files, modem/communication issues, physical access to buildings, escorts needed within facilities, and so on. It is suggested that a training log be developed to document and track information associated with individuals receiving training.

Training Sources

Identify the source or provider of the training. Training may be internal (course developed in-house) or external (contracted to external training agencies).

Dependencies/Constraints/Limitations

Identify all known dependencies, constraints, and/or limitations that could potentially affect training on the project.

Training Resources

This should include hardware/software, instructor availability, training time estimates, projected level of effort, system documentation, and other resources required to familiarize the trainer with the system, produce training materials, and provide the actual training. The identification and availability of other resource groups should also be included.

Vendor Selection

This subsection should describe the criteria for training-vendor selection.

Course Development

This subsection should describe the requirements for internal course development. This should include all related lifecycle activities associated with the development of the training.

Training Environment

Describe the equipment and conditions required for the training, including installations, facilities, and special databases (typically should be a separate, independent development/production environment). Also identify any actions required by other groups, such as trainees, to request training.

Training Materials

Describes the types of training materials required for the training. The training materials developed may include visuals for overhead projectors, handouts, workbooks, manuals, computerized displays, and demonstrations. The training materials and curriculum should accurately reflect the training objective.

Update/Revise Training Plan

Once the training plan is developed, it must be subjected to the same kind of configuration management process as the other system documentation. Training materials should remain current with system enhancements. Describe the change release process with regard to all training documentation.

Chapter **8**

ISO 9001 for Small Projects

INTRODUCTION TO ISO 9001 FOR SMALL PROJECTS

Many organizations are intimidated by the amount of documentation associated with ISO 9001 conformance requirements. The benefit of ISO 9001 implementation in support of larger software development efforts is something that has long been recognized and accepted. Project managers embrace ISO 9001 for larger software development efforts as it provides early notice of project problems. Smaller projects* should also strive for process control and improvement and can benefit from it as well.

The likelihood of small businesses encountering ISO 9001 implementation problems is potentially greater due to:

- Minimal available resources
- Difficulty in understanding and applying the standards
- Costs involved in setting up and maintaining a quality management system

One thing that distinguishes small settings from medium to large settings is that the resources an organization would expect to use to support process improvement efforts (appraisals, consulting, and infrastructure building) would be a large percentage of its operating budget. ISO 9001 Clauses 4.1 and 7.3 require that all criteria and assumptions taken into account in producing the design and development should be carefully considered and written down. This should be used in checks to make sure that there are neither conflicts nor omissions.

To help, ISO sells a book entitled *ISO 9001 for Small Businesses* (ISBN 92-67-10363-6). With only a few people involved, communications in a small business can often be simple and more direct. Individuals are expected to undertake a wide variety of tasks within the business. Decision making is confined to a few people (or even one). The key to implementing ISO 9001 controls on a smaller scale is to

---

*A small project is defined here as a project having 20 or less individuals. Each organization must decide on what it defines as a small project.

have the process documentation in place. Also, instead of having all the documentation relate to a single project, the supporting plans (i.e., software requirements management, configuration management, quality assurance, etc.) describe the processes adopted by the managing organization. In a typical organization, you will find three layers of management: project, program, and organization.

To provide support for the small software project, the policies are developed at the organizational level and the supporting plans at the program level. The project is then required to follow these program-level plans and only write its project-specific documentation. This approach provides several immediate benefits. It allows the technical manager to get down to the business of managing development, encourages standardization across the organization, and encourages participation in process definition activities at the organizational level.

PROJECT MANAGEMENT PLAN—SMALL PROJECTS

The purpose of the small project plan (SPP) is to help project managers during the development of a small software project. This template is useful for projects with time estimates ranging from one person-month to one person-year. It is designed for projects that strive to follow repeatable and defined processes. This project plan is used to refer to existing organizational policies, plans, and procedures. Exceptions to existing policies, plans, and procedures must be documented with additional information appropriate to each project's SPP.

The SPP provides a convenient way to gather and report the critical information necessary for a small project without incurring the overhead of documentation necessary to support the additional communication channels of a larger project. Software quality assurance (SQA) procedures will be captured in Appendix A of the SPP. Appendix B will be used to document software configuration management (SCM) activities. Project management and oversight activities (i.e., risk tracking and problem reporting) will be documented in Appendix C of the SPP. Requirements for the project will be captured in Appendix D of the SPP rather than a separate software requirements specification (SRS). Table 8-1 provides a suggested outline for a SPP.

Small Software Project Management Plan Document Guidance

This small software project management plan template is designed to facilitate the definition of processes and procedures relating to software project management activities. This template was developed using IEEE Standards 12207.0 and 12207.1, Standards for Information Technology—Software Life Cycle Processes [39]; IEEE Std 1058, IEEE Standard for Software Project Management Plans [15]; IEEE Std 730, IEEE Standard for Software Quality Assurance Plans [3]; IEEE Std 828, IEEE Standard for Software Configuration Management Plans [4]; IEEE Std 830, IEEE Recommended Practice for Software Requirements Specifications [6]; IEEE Std

Table 8-1. Small software project management
plan document outline

Title Page
Revision Page
Table of Contents
1. Introduction
    1.1 Identification
    1.2 Scope
    1.3 Document Overview
    1.4 Relationship to Other Plans
2. Acronyms and Definitions
3. References
4. Overview
    4.1 Relationships
    4.2 Source Code
    4.3 Documentation
    4.4 Project Resources
    4.5 Project Constraints
5. Software Process
    5.1 Software Development Process
        5.1.1 Life Cycle Model
    5.2 Software Engineering Activities
        5.2.1 Handling of Critical Requirements
        5.2.2 Recording Rationale
        5.2.3 Computer Hardware Resource Utilization
        5.2.4 Reusable Software
        5.2.5 Software Testing
6. Schedule
Appendix A. Software Quality Assurance
Appendix B. Software Configuration Management
Appendix C. Risk Tracking/Project Oversight
Appendix D. Software Requirements Specification
Requirements Traceability Matrix

1044, Standard Classification for Software Anomalies [13]; IEEE Std 982.1, Standard Dictionary of Measures to Produce Reliable Software [7]; IEEE Std 1045, Standard for Software Productivity Metrics [14]; and IEEE Std 1061, Software Quality Metrics Methodology [16], which have been adapted to support ISO 9001 requirements.

This plan must be supplemented with separate supporting plans. The idea is that this plan may be used to support the development of an SPP for projects operating under the same configuration management, quality assurance, risk management, and so on. policies and procedures. This eliminates redundant documentation, encourages conformity among development efforts, facilitates process improvement through the use of a common approach, and more readily allows for the cross-project transition of personnel.

Introduction

This section should provide a brief summary description of the project and the purpose of this document.

Identification

This subsection should briefly identify the system and software to which this SPP applies. It should include such items as identification number(s), title(s), abbreviations(s), version number(s), and release number(s). It should also outline deliverables, special conditions of delivery, or other restrictions/requirements (e.g., delivery media). An example follows:

> *This document applies to the software development effort in support of the development of [Software Project] Version [xx]. A detailed development timeline in support of this effort is provided in the supporting project Master Schedule.*

Scope

This subsection should briefly state the purpose of the system and software to which this SPP applies. It should describe the general nature of the existing system/software and summarize the history of system development, operation, and maintenance. It should also identify the project sponsor, acquirer, user, developer, support agencies, current and planned operating sites, and relationship of this project to other projects. This overview should not be construed as an official statement of product requirements. It will only provide a brief description of the system/software and will reference detailed product requirements outlined in the SRS, Appendix D. An example follows:

> *[Project Name] was developed for the [Customer Information]. [Brief summary explanation of product.]*
>
> *This overview shall not be construed as an official statement of product requirements. It will only provide a brief description of the system/software and will reference detailed product requirements outlined in the [Project Name] SRS.*

Document Overview

This subsection should summarize the purpose, contents, and any security or privacy issues that should be considered during the development and implementation of the SPP. It should specify the plans for producing scheduled and unscheduled updates to the SPP and methods for disseminating those updates. This subsection should also explain the mechanisms used to place the initial version of the SPP under change control and to control subsequent changes to the SPP. An example is provided:

> *The purpose of the [Project Name] Small Project Plan (SPP) is to guide [Company Name] project management during the development of [Product Identification]. Software Quality Assurance (SQA) procedures are captured in Appendix A of the SPP. Appendix B is used to document Software Configuration Management (SCM) activities where these activities deviate from [Organization Name] SCM Plan. Project management and oversight activities (i.e., risk tracking, problem reporting) are documented in Appendix C of the SPP where these activities deviate from the [Organization Name] Risk Management Plan. Requirements for the project will be captured in the [Project Name] Software Requirements Specification (SRS). This plan will be placed under configuration management controls according the [Organization Name] Software Configuration Management Plan. Updates to this plan will be handled according to relevant SCM procedures and reviews as described in the [Organization Name] Software Quality Assurance Plan.*

Relationship to Other Plans

This subsection should describe the relationship of the SPP to other project management plans. If organizational plans are followed, this should be noted here. An example is provided:

> *There are several other [Organization Name] documents that support the information contained within this plan. These documents include: [Organization Name] Software Configuration Management Plan, [Organization Name] Software Quality Assurance Plan, and [Organization Name] Measurement and Measurements Plan. [Project Name] project-specific documentation relating to this plan include: [Project Name] Software Requirements Specifications and [Project Name] Master Schedule. This plan has been developed in accordance with all [Organization Name] software development processes, policies, and procedures.*

Acronyms and Definitions

This section should identify acronyms and definitions used within the SPP for each project.

References

This section should identify the specific references used within the SPP. Reference to all associated software project documentation, including the statement of work and any amendments should be included.

Overview

This section of the SPP should list the items to be delivered to the customer, delivery dates, delivery locations, and quantities required to satisfy the terms of the contract. This list should not be construed as an official statement of product requirements. It will only provide an outline of the system/software requirements and will reference detailed product requirements outlined in the SRS, Appendix D. An example follows:

> *The project schedule in support of [Project Name] was initiated [Date] with a target completion date of [Date]. Incremental product deliveries may be requested, but none are identified at this time. The delivery requirements are to install [Project Name] on the current [Project Name] external website. [Project Name] will be delivered once deployment has occurred, and [Customer Name] acceptance of the product. [Customer Name] may request the installation of [Product Name] at one additional location. A [Project Name] user's manual shall also be considered to be required as part of this delivery. Detailed schedule guidance is provided by [document name], located [document location]. Detailed requirements information is provided by [document name], located [document location].*

Relationships

This subsection should identify interface requirements and concurrent or critical development efforts that may directly or indirectly affect product development efforts.

Source Code

This subsection should identify source code deliverables. An example is provided:

> *There are no requirements for source code deliverables. If [customer name] requests the source code, it will be delivered on CD-ROM.*

Documentation

This subsection should itemize documentation deliverables and their required format. It should also contain, either directly or by reference, the documentation plan for the software project. This documentation plan may either be a separate document, stating how documentation is going to be developed and delivered, or may be included in this plan as references to existing standards with documentation deliverables and schedule detailed herein. An example follows:

> *[Project Name] will have online help; a user's manual will be created from this online help system. The user's manual will be posted on the [Project Name] website and will be available for download by requesting*

customers. Please refer to the [Project Name] master schedule for a de-velopment timeline of associated user's online help and manual.

Project Resources

This subsection should describe the project's approach by describing the tasks (e.g., req. → design → implementation → test) and efforts (update documentation, etc.) required to successfully complete the project. It should state the nature of each major project function or activity and identify the individuals who are responsible for those functions or activities.

This section should also describe the makeup of the team, project roles, and internal management structure of the project. Diagrams may be used to depict the lines of authority, responsibility, and communication within the project. Figure 8-1 is an example of an organizational chart.

The relationship and interfaces between the project team and all nonproject organizations should also be defined by the SPP. Minimum interfaces include those with the customer, elements (management, SQA, SCM, etc.), and other support teams. Relationships with other contracting agencies or organizations outside the scope of the project that will impact the project require special attention within this section.

Project Constraints

This subsection should describe the limits of the project, including interfaces with other projects, the application of the program's SCM and SQA (including any divergence from those plans), and the relationship with the project's customer. This section should also describe the administrative and managerial boundaries between the project and each of the following entities: parent organization, customer organization, subcontracted organizations, or any organizational entities that interact with the project.

This subsection should capture the anticipated volume of the project through quantifiable measurements such as lines of code, function points, and number of units modified, number of pages of documentation generated or changed. It may be

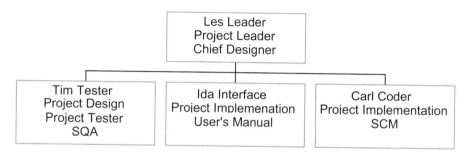

Figure 8-1. Example project organization.

useful, if the project is well defined in advance, to break down project activities and perform size estimates on each individual activity. This information can then be tied to the project schedule.

Software Development Process

This subsection of the SPP should define the relationships among major project functions and activities by specifying the timing of major milestones, baselines, reviews, work products, project deliverables, and signature approvals. The process model may be described using a combination of graphical and textual notations that include project initiation and project termination activities.

Life Cycle Model

This subsection of the SPP should identify the life cycle models used for the software development process. It should describe the project organizational structure, organizational boundaries and interfaces, and individual responsibilities for the various software development elements.

Software Engineering Activities

Handling of Critical Requirements. This subsection should identify the overall software engineering methodologies to be used during requirements elicitation and system design. An example follows:

> *Please refer to the [Organization Name] Software Requirements Management Plan for information regarding the elicitation and management of customer requirements.*

Recording Rationale. This subsection should define the methodologies used to capture design and implementation decisions. An example is provided:

> *All initial design and implementation decisions will be recorded and posted in the [Project Name] portal project workspace. All formal design documentation will follow the format as described in the [Organization Name] Design Documentation Template.*

Computer Hardware Resource Utilization. This subsection should describe the approach to be followed for allocating computer hardware resources and monitoring their utilization. An example text is provided:

> *All computer hardware resource utilization is identified in the [Project Name] Spend Plan. Resource utilization is billed hourly as an associated site contract charge. Twenty thousand dollars has been allocated to support the procurement of a development environment. Charges in support*

of project equipment purchases will be recorded by the Project Manager, [Name], and reported to senior management weekly.

Reusable Software. This subsection should describe the approach for identifying, evaluating, and reporting opportunities to develop reusable software products. The following example is provided:

It is the responsibility of each developer to identify reusable code during the development process. Any item identified will be placed in the [Organization Name] Reuse library that is hosted on the [location]. It is the responsibility of each programmer to determine whether items in the reuse library may be employed during [Project Name] development.

Software Testing. This subsection should be further divided to describe the approach for software implementation and unit testing. See the following:

This project will use existing templates for software test plans and software unit testing when planning for testing. These separate documents will be hosted on the [location].

Schedule

This section of the SPP should be used to capture the project's schedule, including milestones and critical paths. Options include Gantt charts (*Milestones Etc.*™, *Microsoft Project*™), Pert charts, or simple time lines. Figure 8-2 is an example schedule created with *Milestones, Etc.*™ for a short (two month) project.

Appendix A. Software Quality Assurance

This appendix should be divided into the following sections to describe the approach for software quality assurance (SQA): Software quality assurance evaluations, software quality assurance records, independence in software quality assurance, and corrective action. This appendix should reference the organizational- or program-level quality assurance policies, plans, and procedures, and should detail any deviations from the organizational standard. If a separate SQA plan has been created in support of the development work, simply provide a reference to that document as follows:

Refer to the [Organization Name] Software Quality Assurance Plan for all applicable processes and procedures.

Appendix B. Software Configuration Management

This appendix should be divided into the following sections to describe the approach for software configuration management (SCM): configuration identifica-

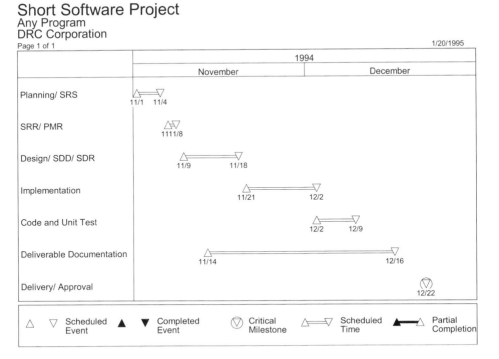

Figure 8-2. Sample project schedule.

tion, configuration control, configuration status accounting, configuration audits, and packaging, storage, handling, and delivery. This appendix should reference the organizational- or program-level configuration management policies, plans, and procedures, and should detail any deviations from the organizational standard. If a separate SCM plan has been created in support of the development work, simply provide a reference to that document as follows:

> *Refer to the [Organization Name] Software Configuration Management Plan for all applicable processes and procedures.*

Appendix C. Risk Tracking/Project Oversight

This appendix should be divided into the following sections to describe the approach for project risk identification and tracking: risk management strategies, measurement activities, and identified project risks. This appendix should reference the organizational- or program-level risk and project management policies, plans, and procedures, and should detail any deviations from the organizational standard. If a separate measurement and measurements plan has been created in support of the development work, simply provide a reference to that document as follows:

The following measures will be taken as part of project oversight activities. Please refer to the [Organization Name] Software Project Measurement Plan for descriptions.

(list of measures)

Appendix D. Software Requirements Specification

This appendix is the equivalent of the software requirements specification (SRS) required for large projects. It is produced early in the project life cycle and contains descriptions of all project requirements. This appendix should reference the organizational- or program-level requirements management policies, plans, and procedures and should detail any deviations from the organizational standard. If a separate SRS has been created in support of the development work, simply provide a reference to that document.

Appendix A

IEEE Standards Abstracts

Standard number	Standard name	Description
IEEE Std 610.12-1990 Reaffirmed Sept 2002	IEEE Standard Glossary of Software Engineering Terminology	Identifies terms currently in use in the field of software engineering. Standard definitions for those terms are established.
IEEE Std 730-2002 Revised Sept 2002	IEEE Standard for Software Quality Assurance Plans	Uniform, minimum acceptable requirements for preparation and content of software quality assurance plans (SQAPs) are provided. This standard applies to the development and maintenance of critical software. For noncritical software, or for software already developed, a subset of the requirements of this standard may be applied.
IEEE Std 828-1998	IEEE Standard for Software Configuration Management Plans	The minimum required contents of a software configuration management plan (SCMP) are established, and the specific activities to be addressed and their requirements for any portion of a software product's life cycle are defined.
IEEE Std 829-1998	IEEE Standard for Software Test Documentation	A set of basic software test documents is described. This standard specifies the form and content of individual test documents. It does not specify the required set of test documents.
IEEE Std 830-1998	IEEE Recommended Practice for Software Requirements Specifications	The content and qualities of a good software requirements specification (SRS) are described and several sample SRS outlines are presented. This recommended practice is aimed at specifying requirements of software to be developed but also can be applied to assist in the selection of in-house and commercial software products.
IEEE Std 982.1-1988 (	IEEE Standard Dictionary of Measures to Produce Reliable Software	This standard provides a set of measures indicative of software reliability that can be applied to the software product as well as to the development and support processes. It was motivated by the needs of software developers and users confronted with a plethora of models, techniques, and measures.

Standard	Title	Description
ANSI/IEEE Std 1008-1987 Reaffirmed Dec. 2002	An American National Standard—IEEE Standard for Software Unit Testing	This standard's primary objective is to specify a standard approach to software unit testing that can be used as a basis for sound software engineering practice.
IEEE Std 1012-1998	IEEE Standard for Software Verification and Validation	Software verification and validation (V&V) processes, which determine whether development products of a given activity conform to the requirements of that activity, and whether the software satisfies its intended use and user needs, are described. This determination may include analysis, evaluation, review, inspection, assessment, and testing of software products and processes. V&V processes assess the software in the context of the system, including the operational environment, hardware, interfacing software, operators, and users.
IEEE Std 1012a-1998	Supplement to IEEE Standard for Software Verification and Validation: Content Map to IEEE/EIA 12207.1-1996	The relationship between the two sets of requirements in plans for verification and validation of software, found in IEEE Std 1012-1998 and IEEE/EIA 12207.1-1996, is explained so that users may produce documents that comply with both standards.
IEEE Std 1016-1998 (Sept 23)	IEEE Recommended Practice for Software Design Descriptions	The necessary information content and recommendations for an organization for software design descriptions (SDDs) are described. An SDD is a representation of a software system that is used as a medium for communicating software design information. This recommended practice is applicable to paper documents, automated databases, design description languages, or other means of description.
IEEE Std 1028-1997 Reaffirmed Sept 2002	IEEE Standard for Software Reviews	This standard defines five types of software reviews, together with procedures required for the execution of each review type. This standard is concerned only with the reviews; it does not define procedures for determining the necessity of a review, nor does it specify the disposition of the results of the review. Review types include management reviews, technical reviews, inspections, walk-throughs, and audits.

(continued)

249

Standard number	Standard name	Description
IEEE Std 1042-1987	IEEE Guide to Software Configuration Management	This guide describes the application of configuration management disciplines to management of software engineering projects. This guide serves three groups: developers of software, the software management community, and those responsible for preparation of SCM plans. Software configuration management consists of two major aspects: planning and implementation. This guide focuses on software configuration management planning and provides broad perspectives for the understanding of software configuration management.
IEEE Std 1044-1993 Reaffirmed Sept 2002	IEEE Standard Classification for Software Anomalies	A uniform approach to the classification of anomalies found in software and its documentation is provided. The processing of anomalies discovered during any software life cycle phase are described, and comprehensive lists of software anomaly classifications and related data items that are helpful to identify and track anomalies are provided.
IEEE Std 1045-1992 Reaffirmed Dec 2002	IEEE Standard for Software Productivity Metrics	Provides consistent ways to measure the elements that go into computing software productivity are defined. Software productivity metrics terminology is given to ensure an understanding of measurement data for both source code and document production.
IEEE Std 1058-1998	IEEE Standard for Software Project Management Plans	The format and contents of software project management plans, applicable to any type or size of software project, are described. The elements that should appear in all software project management plans are identified.
IEEE Std 1061-1998	IEEE Standard for a Software Quality Metrics Methodology	A methodology for establishing quality requirements and identifying, implementing, analyzing, and validating the process and product software quality metrics is defined. The methodology spans the entire software life cycle.

IEEE Std 1062-1998 Reaffirmed Sept 2002	IEEE Recommended Practice for Software Acquisition	A set of useful quality practices that can be selected and applied during one or more steps in a software acquisition process is described. This recommended practice can be applied to software that runs on any computer system regardless of the size, complexity, or criticality of the software, but is more suited for use on modified off-the-shelf software and fully developed software.
IEEE Std 1063-2001	IEEE Standard for Software User Documentation	Two factors motivated the development of this standard: the concern of the software user communities over the poor quality of much user documentation, and a need for requirements expressed by producers of documentation.
IEEE Std 1074-1997	IEEE Standard for Developing Software Life Cycle Processes	A process for creating a software life cycle process is provided. Although this standard is directed primarily at the process architect, it is useful to any organization that is responsible for managing and performing software projects.
IEEE Std 1175-1991	IEEE Standard Reference Model for Computing System Tool Interconnections	The purpose is to establish agreements for information transfer among tools in the contexts of human organization, a computer system platform, and a software development application. Interconnections that must be considered when buying, building, testing, or using computing system tools for specifying behavioral descriptions or requirements of system and software products are described.
IEEE Std 1175.1-2002	IEEE Guide for CA Software Engineering Tool Interconnections— Classification and Description	Introduces and characterizes the problem of interconnecting Computer-Aided (CA) Software Engineering tools with their environment.
IEEE Std 1219-1998	IEEE Standard for Software Maintenance	The process for managing and executing software maintenance activities is described.

Standard number	Standard name	Description
IEEE Std 1220-1998	IEEE Standard for the Application and Management of the Systems Engineering Process	The interdisciplinary tasks, which are required throughout a system's life cycle to transform customer needs, requirements, and constraints into a system solution, are defined. In addition, the requirements for the systems engineering process and its application throughout the product life cycle are specified. The focus of this standard is on engineering activities necessary to guide product development while ensuring that the product is properly designed to make it affordable to produce, own, operate, maintain, and, eventually, dispose of, without undue risk to health or the environment.
IEEE Std 1228-1994 (Mar 17) Reaffirmed	IEEE Standard for Software Safety Plans	The minimum acceptable requirements for the content of a software safety plan are established. This standard applies to the software safety plan used for the development, procurement, maintenance, and retirement of safety-critical software. This standard requires that the plan be prepared within the context of the system safety program. Only the safety aspects of the software are included. This standard does not contain special provisions required for software used in distributed systems or in parallel processors.
IEEE Std 1233, 1998 Reaffirmed Sept 2002	IEEE Guide for Developing System Requirements Specifications	Guidance for the development of the set of requirements [System Requirements Specification (SyRS)] that will satisfy an expressed need is provided. Developing a SyRS includes the identification, organization, presentation, and modification of the requirements. Also addressed are the conditions for incorporating operational concepts, design constraints, and design configuration requirements into the specification. This guide also covers the necessary characteristics and qualities of individual requirements and the set of all requirements.

Standard	Description
IEEE Std 1320.1-1998 IEEE Standard for Functional Modeling Language—Syntax and Semantics for IDEF0	IDEF0 function modeling is designed to represent the decisions, actions, and activities of an existing or prospective organization or system. IDEF0 may be used to model a wide variety of systems, composed of people, machines, materials, computers, and information of all varieties, and structured by the relationships among them, both automated and nonautomated. As the basis of this architecture, IDEF0 may then be used to design an implementation that meets these requirements and performs these functions.
IEEE Std 1320.2-1998 EEE Std 1320.2a IEEE Standard for Conceptual Modeling Language Syntax and Semantics for IDEF1X 97 (IDEF object)	IDEF1X 97 consists of two conceptual modeling languages. The key-style language supports data/information modeling and is downward compatible with the U.S. government's 1993 standard, FIPS PUB 184. The identity-style language is based on the object model with declarative rules and constraints.
IEEE Std 1362-1998 IEEE Guide for Information Technology—System Definition—Concept of Operations (ConOps) Document	The format and contents of a concept of operations (ConOps) document are described. A ConOps is a user-oriented document that describes system characteristics for a proposed system from the users' viewpoint. The ConOps document is used to communicate overall quantitative and qualitative system characteristics to the user, buyer, developer, and other organizational elements (for example, training, facilities, staffing, and maintenance). It is used to describe the user organization(s), mission(s), and organizational objectives from an integrated systems point of view.
IEEE Std 1420.1-1995 Reaffirmed June 2002 IEEE Standard for IT Software Reuse—Data Model for Reuse Library Interoperability: Basic Interoperability Data Model (BIDM)	The minimal set of information about assets that reuse libraries that should be able to be exchanged to support interoperability is provided.

(continued)

253

Standard number	Standard name	Description
IEEE Std 1420.1a-1996 Reaffirmed June 2002	Supplement to IEEE Standard for IT Software Reuse—Data Model for Reuse Library Interoperability: Asset Certification Framework	A consistent structure for describing a reuse library's asset certification policy in terms of an asset certification framework is defined, along with a standard interoperability data model for interchange of asset certification information.
IEEE Std 1420.1b-1999 Reaffirmed June 2002	IEEE Trial-Use Supplement to IEEE Standard for IT Software Reuse—Data Model for Reuse Library Interoperability: Intellectual Property Rights Framework	This extension to the Basic Interoperability Data Model (IEEE Std 1420.1-1995) incorporates intellectual property rights issues into software asset descriptions for reuse library interoperability.
IEEE Std 1462-1998	IEEE Standard—Adoption of ISO/IEC 14102: 1995—Information Technology—Guideline for the Evaluation and Selection of CASoftware Engineering Tools	Deals with the evaluation and selection of CASoftware Engineering tools, covering a partial or full portion of the software engineering life cycle. The adoption of the international standard by IEEE includes an implementation note, which explains terminology differences, identifies related IEEE standards, and provides interpretation of the international standard.

IEEE Std 1465-1998	IEEE Standard—Adoption of International Standard ISO/IEC 12119: 1994(E)—Information Technology—Software packages—Quality Requirements and Testing	Quality requirements for software packages and instructions on how to test a software package against these requirements are established. The requirements apply to software packages as they are offered and delivered, not to the production process (including activities and intermediate products, such as specifications).
IEEE Std 1471-2000	IEEE Recommended Practice for Architectural Description of Software Intensive Systems	This recommended practice addresses the activities of the creation, analysis, and sustainment of architectural descriptions. A conceptual framework for architectural description is established. The content of an architectural description is defined. Annexes provide the rationale for key concepts and terminology, the relationships to other standards, and examples of usage.
IEEE Std 1490-1998	IEEE Guide—Adoption of PMI Standard—A Guide to the Project Management Body of Knowledge	The subset of the Project Management Body of Knowledge that is generally accepted is identified and described in this guide. "Generally accepted" means that the knowledge and practices described are applicable to most projects most of the time, and that there is widespread consensus about their value and usefulness. It does not mean that the knowledge and practices should be applied uniformly to all projects without considering whether they are appropriate.
IEEE Std 1517-1999	IEEE Standard for Information Technology—Software Life Cycle Processes—Reuse Processes	A common framework for extending the software life cycle processes of IEEE/EIA Std 12207.0-1996 to include the systematic practice of software reuse is provided. This standard specifies the processes, activities, and tasks to be applied during each phase of the software life cycle to enable a software product to be constructed from reusable assets. It also specifies the processes, activities, and tasks to enable the identification, construction, maintenance, and management of assets supplied.

(continued)

Standard number	Standard name	Description
IEEE Std 1540-2001	IEEE Standard for Software Life Cycle Processes—Risk Management	A process for the management of risk in the life cycle of software is defined. It can be added to the existing set of software life cycle processes defined by the IEEE/EIA 12207 series of standards, or it can be used independently.
IEEE Std 2001-2002	IEEE Recommended Practice for Internet Practices—Web Page Engineering—Intranet/Extranet Applications	This standard defines recommended practices for Web page engineering. It addresses the needs of webmasters and managers to effectively develop and manage World Wide Web projects (internally via an intranet or in relation to specific communities via an extranet).
IEEE/EIA 12207.0-1996	Standard for Information Technology—Software Life Cycle Processes—Software Life Cycle Processes	ISO/IEC 12207 provides a common framework for developing and managing software. IEEE/EIA 12207.0 consists of the clarifications, additions, and changes accepted by the Institute of Electrical and Electronics Engineers (IEEE) and the Electronic Industries Association (EIA) as formulated by a joint project of the two organizations.
IEEE/EIA 12207.1-1996	Standard for Information Technology—Software Life Cycle Processes—Software Life Cycle Processes—Life Cycle Data	ISO/IEC 12207 provides a common framework for developing and managing software. IEEE/EIA 12207.0 consists of the clarifications, additions, and changes accepted by the Institute of Electrical and Electronics Engineers (IEEE) and the Electronic Industries Association (EIA) as formulated by a joint project of the two organizations. IEEE/EIA 12207.1 provides guidance for recording life cycle data resulting from the life cycle processes of IEEE/EIA 12207.0.

| IEEE/EIA 12207.2-1997 | Standard for Information Technology—Software Life Cycle Processes—Software Life Cycle Processes—Implementation Considerations | ISO/IEC 12207 provides a common framework for developing and managing software. IEEE/EIA 12207.0 consists of the clarifications, additions, and changes accepted by the Institute of Electrical and Electronics Engineers (IEEE) and the Electronic Industries Association (EIA) as formulated by a joint project of the two organizations. IEEE/EIA 12207.2 provides implementation consideration guidance for the normative clauses of IEEE/EIA 12207.0. The guidance is based on software industry experience with the life cycle processes presented in IEEE/EIA 12207.0. |
| IEEE Std 14143.1-2000 | IT—Software Measurement—Functional Size Measurement—Part 1: Definition of Concepts | Implementation notes that relate to the IEEE interpretation of ISO/IEC 14143-1:1998 are described. |

Appendix **B**

Comparison of ISO 9001 to IEEE Standards

Software Life Cycle Process	ISO 9001 Clauses	IEEE Standards
5. Primary Life Cycle Processes	4.1, General Requirements 7.1, Planning of Product Realization	1074, Developing a Software Project Life Cycle Process
5.1 Acquisition	7.2.2, Review of Requirements Related to the Product 7.4.1, Purchasing Process 7.4.2, Purchasing Information 7.4.3, Verification of Purchased Product	829, Software Test Documentation 1062, Software Acquisition 1362, Concept of Operations
5.2 Supply	7.1, Planning of Product Realization 7.2.1, Determination of Requirements Related to the Product 7.2.2, Review of Requirements Related to the Product 7.2.3, Customer Communication	1028, Software Reviews 1058, Software Project Management Plans 1062, Software Acquisition 1220, Application and Management of the Systems Engineering Process 12207.0, Life Cycle Processes
5.3 Development	7.3, Design And Development 7.3.1, Design and Development Planning	829, Software Test Documentation 830, Software Requirements Specifications 1008, Software Unit Testing

(continued)

Practical Support for ISO 9001 Software Project Documentation. By S. Land and J. Walz

Software Life Cycle Process	ISO 9001 Clauses	IEEE Standards
5.3 Development (*cont.*)	7.3.2, Design and Development Inputs 7.3.3, Design and Development Outputs 7.3.4, Design and Development Review 7.3.5, Design and Development Verification 7.3.6, Design and Development Validation 7.3.7, Control of Design and Development Changes	1012, Software Verification and Validation Plans 1016, Software Design Descriptions 1063, Software User Documentation 1074, Developing a Software Project Life Cycle Process 1220, Application and Management of the Systems Engineering Process 1233, Developing System Requirements Specifications 1320.1, 1320.2, Functional Modeling Language—Syntax and Semantics for IDEF0, IDEF1X97 1420.1, 1420.1a, and 1420.1b, Software Reuse—Data Model for Reuse Library Interoperability 1471, Architectural Description of Software Intensive Systems 12207.0, Software Life Cycle Processes 12207.1, Software Life Cycle Processes—Life Cycle Data
5.3.1 Process Implementation	7.1, Planning of Product Realization 7.3.1, Design and Development Planning	1220, Application and Management of the Systems Engineering Process 1074, Developing a Software Project Life Cycle Process 12207.1, Software Life Cycle Processes—Life Cycle Data
5.3.2 System Requirements	7.3.2, Design and Development Inputs	1220, Application and Management of the Systems Engineering Process 1233, System Requirements Specifications

Software Life Cycle Process	ISO 9001 Clauses	IEEE Standards
5.3.2 System Requirements (*cont.*)		1320.1, 1320.2, Functional Modeling Language—Syntax and Semantics for IDEF0, IDEF1X97 12207.1, Software Life Cycle Processes—Life Cycle Data
5.3.3 System Architectural Design	7.3.2, Design and Development Inputs	1233, System Requirements Specifications 1320.1, 1320.2, Functional Modeling Language—Syntax and Semantics for IDEF0, IDEF1X97 12207.1, Software Life Cycle Processes—Life Cycle Data
5.3.4 Software Requirements	7.3.2, Design and Development Inputs	830, Software Requirements Specifications 12207.1, Software Life Cycle Processes—Life Cycle Data
5.3.5 Software Architectural Design	7.3.3, Design and Development Outputs	829, Software Test Documentation 1016, Software Design Descriptions 1063, Software User Documentation 1320.1, 1320.2, Functional Modeling Language—Syntax and Semantics for IDEF0, IDEF1X97 1420.1, 1420.1a, and 1420.1b, Software Reuse—Data Model for Reuse Library Interoperability 1471, Architectural Description of Software Intensive Systems 12207.1, Software Life Cycle Processes—Life Cycle Data
5.3.6 Software Detailed Design	7.3.3, Design and Development Outputs	829, Software Test Documentation 1016, Software Design Descriptions (*continued*)

Software Life Cycle Process	ISO 9001 Clauses	IEEE Standards
5.3.6 Software Detailed Design (*cont.*)		1063, Software User Documentation 12207.1, Software Life Cycle Processes—Life Cycle Data
5.3.7 Software Coding and Testing	7.3.3, Design and Development Outputs	829, Software Test Documentation 1008, Software Unit Testing 1063, Software User Documentation 12207.1, Software Life Cycle Processes—Life Cycle Data
5.3.8 Software Integration	7.3.5, Design and Development Verification	829, Software Test Documentation 1012, Software Verification and Validation Plans 1063, Software User Documentation 12207.1, Software Life Cycle Processes—Life Cycle Data
5.3.9 Software Qualification Testing	7.3.6, Design and Development Validation	829, Software Test Documentation 1012, Software Verification and Validation Plans 1063, Software User Documentation 12207.1, Software Life Cycle Processes—Life Cycle Data
5.3.10 System Integration	7.3.5, Design and Development Verification	829, Software Test Documentation 1012, Software Verification and Validation Plans 12207.1, Software Life Cycle Processes—Life Cycle Data
5.3.11 System Qualification Testing	7.3.6, Design and Development Validation	829, Software Test Documentation 1012, Software Verification and Validation Plans 12207.1, Software Life Cycle Processes—Life Cycle Data

Software Life Cycle Process	ISO 9001 Clauses	IEEE Standards
5.3.12 Software Installation	7.3.6, Design and Development Validation 7.5.1, Control of Production and Service Provision 7.5.5, Preservation of Product	829, Software Test Documentation 1012, Software Verification and Validation Plans 12207.1, Software Life Cycle Processes—Life Cycle Data
5.3.13 Software Acceptance Support	7.3.6, Design and Development Validation 7.5.1, Control of Production and Service Provision	829, Software Test Documentation 1012, Software Verification and Validation Plans 12207.1, Software Life Cycle Processes—Life Cycle Data
5.4 Operation	7.2.3, Customer Communication During Operations and Maintenance 7.5.1, Control of Production and Service Provision 8.2.1, Customer Satisfaction	1063, Software User Documentation 1465, Software Packages—Quality Requirements and Testing
5.5 Maintenance	7.2.3, Customer Communication During Operations and Maintenance 7.3.7, Control of Design and Development Changes 7.5.1, Control of Production and Service Provision	1219, Software Maintenance 12207.0, Software Life Cycle Processes
6. Supporting Processes	4.2, Documentation Requirements 7.1, Planning of Product Realization	
6.1 Documentation	4.2.1, General Documentation Requirements 4.2.2, Quality Manual 4.2.3, Control of Documents 4.2.4, Control of Records	1063, Software User Documentation 12207.1, Software Life Cycle Processes—Life Cycle Data

(*continued*)

Software Life Cycle Process	ISO 9001 Clauses	IEEE Standards
6.1 Documentation (*cont.*)	7.1, Planning of Product Realization 7.3.7, Control of Design and Development Changes	
6.2 Configuration Management	7.1, Planning of Product Realization 7.3.1, Design and Development Planning 7.3.7, Control of Design and Development Changes 7.5.1, Control of Production and Service Provision 7.5.3, Identification and Traceability 7.5.4, Customer Property 7.6, Control of Measuring and Monitoring Device 8.3, Nonconforming Product or Erroneous Configuration Item 8.5.2, Corrective Action	828, Software Configuration Management Plans
6.3 Quality Assurance	5.4.2, Quality Management System Planning 7.3.1, Design and Development Planning 8.2.2, Internal Audit	730, Software Quality Assurance Plans 1061, Software Quality Metrics Methodology 1465, Software Packages— Quality Requirements and Testing 90003, Guidelines for the Application of ISO 9001:2000 to Computer Software
6.4 Verification	7.1, Planning of Product Realization 7.2.2, Review of Requirements Related to the Product 7.3.1, Design and Development Planning	1012, Software Verification and Validation 1028, Software Reviews— Inspections 1044, Classification for Software Anomalies

Software Life Cycle Process	ISO 9001 Clauses	IEEE Standards
6.4 Verification (*cont.*)	7.3.3, Design and Development Outputs 7.3.5, Design and Development Verification 7.3.7, Control of Design and Development Changes 7.4.3, Verification of Purchased Product 7.6, Control of Measuring and Monitoring Devices 8.2.2, Internal Audit 8.2.4, Monitoring and Measurement of Product 8.3, Control of Nonconforming Product	
6.5 Validation	7.1, Planning of Product Realization 7.3.1, Design and Development Planning 7.3.6, Design and Development Validation 7.3.7, Control of Design and Development Changes 7.5.1, Control of Production and Service Provision 7.5.2, Validation of Processes for Production and Service Provision	829, Software Test Documentation 1008, Software Unit Testing 1012, Software Verification and Validation 1028, Software Reviews 1044, Classification for Software Anomalies 12207.1, Software Life Cycle Processes—Life Cycle Data
6.6 Joint Review	7.1, Planning of Product Realization 7.2.2, Review of Requirements Related to the Product 7.2.3, Customer Communication 7.3.1, Design and Development Planning 7.3.4, Design and Development Review	1028, Software Reviews 1044, Classification for Software Anomalies

(continued)

Software Life Cycle Process	ISO 9001 Clauses	IEEE Standards
6.7 Audit	5.6.2, Management Review Inputs 7.3.6, Design and Development Validation 7.4.1, Purchasing Process 8.2.2, Internal Audit 8.5.1, Continual Improvement	1028, Software Reviews 1061, Software Quality Metrics Methodology 15939, Software Measurement Process
6.8 Problem Resolution	7.1, Planning of Product Realization 7.3.1, Design and Development Planning 7.3.4, Design and Development Review 7.5.1, Control of Production and Service Provision 8.3, Control of Nonconforming Product 8.5.2, Corrective Action	1044, Classification for Software Anomalies 1490, Project Management Body of Knowledge (PMBOK) 1540, Risk Management Requirements
7. Organizational Life Cycle Processes	5, Management Responsibility 6, Resource Management 8, Measurement, and Analysis and Improvement	
7.1 Management	4.1, General Requirements 5.1, Management Commitment 5.2, Customer Focus 5.3, Quality Policy 5.4.1, Quality Objectives 5.4.2, Quality Management System Planning 5.5, Responsibility, Authority, and Communication 5.6, Management Review 6.1, Provision of Resources 7.1, Planning of Product Realization	830, Software Requirements Specification 1058, Software Project Management Plans 1490, Project Management Body of Knowledge 12207.0, Software Life Cycle Processes

Software Life Cycle Process	ISO 9001 Clauses	IEEE Standards
7.1 Management (*cont.*)	7.3.1, Design and Development Planning	
7.1 Infrastructure	6.3, Infrastructure	1074, Developing Software Life Cycle Processes 1175.1, Computer-Aided Software Engineering Tool Interconnections—Classification and Description 1462, Evaluation and Selection of Computer-Aided Software Engineering Tools 12207.0, Software Life Cycle Processes
7.2 Improvement	4.1, General Requirements 5.6.1, Management Review 8.1, Measurement, Analysis, and Improvement—General 8.2.2, Internal Audit 8.2.3, Monitoring and Measurement of Processes 8.4, Analysis of Data 8.5.1, Continual Improvement 8.5.3, Preventive Action	730, Software Quality Assurance Plans 1028, Software Reviews 1074, Developing Software Life Cycle Processes 12207.0, Software Life Cycle Processes
7.3 Training	6.2.1, Human Resources—General 6.2.2, Competence, Awareness, and Training	12207.0, Software Life Cycle Processes 12207.1, Software Life Cycle Processes—Life Cycle Data 12207.2, Software Life Cycle Processes—Implementation Considerations

Appendix **C**

Work Products

ACQUISITION

Make/Buy Decision Matrix

This matrix may be used to support the analysis of alternative solutions. This matrix should include a listing of all project requirements and their estimated cost. It should also include a list of all candidate applications that come closest to meeting the specifications as well as a list of reusable components. This matrix should be used to compare and analyze key functions and cost, addressing the qualitative as well as quantitative issues surrounding the selection process. An example decision matrix is provided on the companion CD-ROM as and is entitled *Decision Matrix.xls*.

Table C-1, provides some sample questions to support the analysis of each of the make/buy/mine/commission options.

Alternative Solution Screening Criteria Matrix

An alternative solution screening criteria matrix may be used to support the analysis of alternative solutions. This matrix should include a listing of all project requirements and their estimated cost. It should also include a list of all candidate applications that come closest to meeting the specifications as well as a list of reusable components. This matrix should be used to compare and analyze key functions and cost, addressing the qualitative as well as quantitative issues surrounding the selection process. An example decision matrix is provided on the companion CD-ROM as and is entitled *Decision Matrix.xls*. Suggestions in support of effective screening practices are provided in Table C-2.

Cost–Benefit Ratio

A cost–benefit ratio is a calculation that depicts the total financial return for each dollar invested. Many projects have ready access to staff "hours" rather than accounting figures, which is good enough for decision analysis and resolution. The simplified example in Table C-3 is based on a fictitious case of developing a Web-

Table C-1. Questions in support of make/buy/mine/commission options [95]

Sample questions for the "make" option:
- Are developers with appropriate expertise available?
- How would the developers be utilized if not on this project?
- If additional personnel need to be hired, will they be available within the needed time frame?
- How successful has the organization been in developing similar assets?
- What specific flexibilities are gained by developing products in-house as opposed to purchasing them?
- What development tools and environments are available? Are they suitable? How skilled is the targeted workforce in their use?
- What are the costs of development tools and training, if needed?
- What are the other specific costs of developing the asset in-house?
- What are the other specific benefits of developing the asset in-house?

Sample questions for the "buy" option:
- What assets are commercially available?
- How well does the COTS software conform to the product-line architecture?
- How closely does the available COTS software satisfy the product-line requirements?
- Are small changes in COTS software a viable option?
- Is source code available with the software? What documentation comes with the software?
- What are the integration challenges?
- What rights to redistribute are purchased with the COTS software?
- What are the other specific costs associated with purchasing the software?
- What are the other specific benefits associated with purchasing the software?
- How stable is the vendor?
- How often are upgrades produced? How relevant are the upgrades to the product line?
- How responsive is the vendor to user requests for improvements?
- How strong is the vendor support?

Sample questions for the "mine" option:
- What legacy systems are available from which to mine assets?
- How close is the functionality of the legacy software to the functionality that is required?
- What is the defect track record for the software and nonsoftware assets?
- How well is the legacy system documented?
- What mining strategies are appropriate?
- How expensive are those strategies?
- What experience does the organization have in mining assets?
- What mining tools are available? Are they appropriate? How skilled is the workforce in their use?
- What are the costs of mining tools and training, if needed?
- What are the other specific costs associated with mining the asset?
- What are the other specific benefits associated with mining the asset?
- What changes need to be made to the legacy asset to perform the mining?
- What are the costs and risks of those changes?
- What types of noncode assets are available?
- What modifications or additions need to be made to them?

Table C-1. *Continued*

Sample questions for the "commission" option:
- What contractors are available to develop the asset?
- What is the track record of the contractor in terms of schedule and budget?
- Is the acquiring organization skilled in supervising contracted work?
- Are the requirements defined to the extent that the asset can be subcontracted?
- Are interface specifications well defined and stable?
- What experience does the contractor have with the principles of product-line development?
- Who needs to own the asset? Who maintains it?
- What are the other specific costs associated with commissioning the asset?
- What are the other specific benefits associated with commissioning the asset?
- What are the costs of maintaining the commissioned asset?
- Does commissioning the asset involve divulging to the contractor any technology or information that it is in the acquiring organization's interest to keep in-house?

Table C-2. Tips for effective solution screening

Strike a reasoned balance between efficiency and completeness	Efficient screening often requires employing high-level criteria that can quickly exclude many COTS components. For example, a common and often appropriate screening strategy suggests that examining only the top few (e.g., 3–5) competitive components in terms of sales or customer-installed base. However, there are often reasons to additionally consider smaller, niche players that provide a unique or tailored capability more in line with system expectations. This suggests that there may be multiple pathways for inclusion of a component.
Avoid premature focus on detailed architecture and design	Although high-level architecture is an appropriate screening criterion, there is a tendency to prematurely focus on detailed architecture and design as screening criteria (for example, specifying the architectural characteristics unique to one component). When architectural or design decisions have already been made that must be reflected by the chosen component, then it is important that criteria reflect these decisions. However, to fight against the tendency to think within the box, consider the risks and potential workarounds if a highly similar capability is delivered in a different manner.
Optimize the order in which you apply screening criteria	Criteria will vary along dimensions of ability to discriminate and effort required to obtain data. By their nature, some criteria will be highly discriminatory in determining which components are appropriate. Other criteria, though important, will eliminate fewer components from consideration. Still other criteria will require more work than the norm to accumulate data. Consider both how discriminating a criterion is and the effort needed to evaluate components against the criterion when determining an order for applying screening criteria.
Market share/growth are good screens for long-lived systems	A defining characteristic of many systems is expected lifespan. Often, systems are used for over 20 years. COTS components in these systems should come from financially sound companies with good prospects to remain in business.

based user-support program that is estimated to cost 1000 hours to develop and deliver, and which should save $6400 staff hours in the first year due to a reduction in field service calls.

SUPPLY

Recommendations for Software Acquisition

IEEE Std 1062, IEEE Recommended Practice for Software Acquisition describes the software acquisition process. IEEE Std 12207.0 also recommends a set of objectives in support of the Acquisition process. Table C-4 describes these combined recommendations and should be used to support the complete life cycle of the acquisition process.

Table C-5 lists the acquisition checklists provided in IEEE Std 1062. Items listed also support conformance with IEEE 12207.0. Items in boldface font are included in this appendix.

Organizational Acquisition Strategy Checklist

IEEE Std 1062, IEEE Recommended Practice for Software Acquisition, provides a example checklist in support of the definition of an organizational acquisition strategy as described in Table C-6. Any checklist used by an organization to determine acquisition strategy should reflect the requirements of the acquiring organization. An electronic version of this document is provided on the companion CD-ROM, entitled *Acquisition Strategy Checklist.doc.*

Supplier Evaluation Criteria

IEEE Std 1062, IEEE Recommended Practice for Software Acquisition, provides a checklist in support supplier evaluation. This evaluation criterion is provided in Table C-7. The information provided here provides an illustrative example and should be tailored to reflect organizational process needs. An electronic version of this checklist is provided on the companion CD-ROM, entitled *Supplier Checklist.doc.*

Table C-3. Cost–benefit ratio calculation

	Benefit (first year hours)	Total investment (hours)	Cost–benefit ratio	Return after 1 year. Each dollar invested returned over:
Planned	6,400	1,000	6.4	$6
Actual	6,000	800	7.5	$7

Table C-4. Recommendations for Software Acquisition Process

Rec. 1	Planning organizational strategy. Review acquirer's objectives and develop a strategy for acquiring software. [17]
Rec. 2	Implementing the organization's process. Establish a software acquisition process that fits the organization's needs for obtaining a quality software product. Include appropriate contracting practices. [17]
Rec. 3	Determining the software requirements. Define the software being acquired and prepare quality and maintenance plans for accepting software supplied by the supplier. [17]
Rec. 4	Identifying potential suppliers. Select potential candidates who will provide documentation for their software, demonstrate their software, and provide formal proposals. Failure to perform any of these actions is basis to reject a potential supplier. Review supplier performance data from previous contracts. [17]
Rec. 5	Preparing contract requirements. Describe the quality of the work to be done in terms of acceptable performance and acceptance criteria, and prepare contract provisions that tie payments to deliverables. [17] Develop a contract that clearly expresses the expectations, responsibilities, and liabilities of both the acquirer and the supplier. Review the contract with legal counsel. [39]
Rec. 6	Evaluating proposals and selecting the supplier. Evaluate supplier proposals, select a qualified supplier, and negotiate the contract. Negotiate with an alternate supplier, if necessary. [17] Obtain products and /or services that satisfy the customer's need. Qualify potential suppliers through an assessment of their capability to perform the required software. [39]
Rec. 7	Managing supplier performance. Monitor the supplier's progress to ensure that all milestones are met and to approve work segments. Provide all acquirer deliverables to the supplier when required. [17] Manage the acquisition so that specified constraints and goals are met. Establish a statement of work to be performed under contract. Regularly exchange progress information with the supplier. [39]
Rec. 8	Accepting the software. Perform adequate testing and establish a process for certifying that all discrepancies have been corrected and that all acceptance criteria have been satisfied. [17], [39]
Rec. 9	Using the software. Conduct a follow-up analysis of the software acquisition contract to evaluate contracting practices, record lessons learned, and evaluate user satisfaction with the product. Retain supplier performance data. [17] Establish a means by which the acquirer will assume responsibility for the acquired software product or service. [39]

Supplier Performance Standards

IEEE Std 1062, IEEE Recommended Practice for Software Acquisition, provides a checklist in support of determining satisfactory supplier performance. This evaluation, provided in Table C-8, is provided as an example and should be based upon all known requirements and constraints unique to the development effort. An electron-

Table C-5. IEEE Std 1062-1998 acquisition
checklist summary

Checklist 1:	**Organizational strategy**
Checklist 2:	Define the software
Checklist 3:	**Supplier evaluation**
Checklist 4:	Supplier and acquirer obligations
Checklist 5:	Quality and maintenance plans
Checklist 6:	User survey
Checklist 7:	**Supplier performance standards**
Checklist 8:	Contract payments
Checklist 9:	Monitor supplier progress
Checklist 10:	Software evaluation
Checklist 11:	Software test
Checklist 12:	Software acceptance

ic version of this checklist is provided on the companion CD-ROM, identified as
Supplier Performance Standards.doc.

DEVELOPMENT

Requirements Traceability

Requirements traceability is an ISO 9001 requirement. This traceability throughout
the defined life cycle can be accomplished by the addition of a traceability matrix.
Table C-9 supports backward and forward traceability for validation testing. Additional columns should be added to support traceability to software, or system, design and development. The conversion of this type of matrix to a database tracking
system is a common practice for developing organizations. There are many commercially available tools that support requirements tracking.

Software Development Standards Description

The definition of standards associated with the development of software is integral
to the success of any software development effort. The following outline is provided by section 6.17 of IEEE Std 12207.1, Software Life Cycle Processes Life Cycle
Data [39].

a) Generic description information

Date of issue and status

Scope

Issuing organization

References

Context

Notation for description

Body

Summary

Glossary

Change history

Table C-6. Acquisition strategy checklist [17]

1) Who will provide software support?	Supplier ☐	Acquirer ☐
2) Is maintenance documentation necessary?	Yes ☐	No ☐
3) Will user training be provided by the supplier?	Yes ☐	No ☐
4) Will acquirer's personnel need training?	Yes ☐	No ☐
5) When software conversion or modification is planned:		
a) Will supplier manuals sufficiently describe the supplier's software?	Yes ☐	No ☐
b) Will specification be necessary to describe the conversion or modification requirements and the implementation details of the conversion or modification?	Yes ☐	No ☐
c) Who will provide these specifications?	Supplier ☐	Acquirer ☐
d) Who should approve these specifications? _____		
6) Will source code be provided by the supplier so that modifications can be made?	Yes ☐	No ☐
7) Are supplier publications suitable for end users?	Yes ☐	No ☐
a) Will unique publications be necessary?	Yes ☐	No ☐
b) Will unique publications require formal acceptance?	Yes ☐	No ☐
c) Are there copyright or royalty issues?	Yes ☐	No ☐
8) Will the software be evaluated and certified?	Yes ☐	No ☐
a) Is a survey of the supplier's existing customers sufficient?	Yes ☐	No ☐
b) Are reviews and audits desirable?	Yes ☐	No ☐
c) Is a testing period preferable to demonstrate that the software and its associated documentation are usable in their intended environment?	Yes ☐	No ☐
d) Where will the testing be performed? _____		
e) Who will perform the testing? _____		
f) When will the software be ready for acceptance? _____		
9) Will supplier support be necessary during initial installations of the software by the end users?	Yes ☐	No ☐
10) Will subsequent releases of the software be made?	Yes ☐	No ☐
a) If so, how many? _____ Compatible upgrades?	Yes ☐	No ☐
11) Will the acquired software require rework whenever operating system changes occur?	Yes ☐	No ☐
a) If so, how will the rework be accomplished? _____		
12) Will the acquired software commit the acquiring organization to a software product that could possibly be discontinued?	Yes ☐	No ☐
13) What are the risks/options if software is not required? _____		

Table C-7. Supplier evaluation checklist [17]

Financial soundness
  1) Can a current financial statement be obtained for examination?
  2) Is an independent financial rating available?
  3) Has the company or any of its principals ever been involved in bankruptcy or litigation?
  4) How long has the company been in business?
  5) What is the company's history?

Experience and capabilities
  1) On a separate page, list by job function the number of people in the company.
  2) On a separate page, list the names of sales and technical representatives and support personnel. Can they be interviewed?
  3) List the supplier's software products that are sold and the number of installations of each.
  4) Is a list of users available?

Development and control processes
  1) Are software development practices and standards used?
  2) Are software development practices and standards adequate?
  3) Are the currently used practices written down?
  4) Are documentation guidelines available?
  5) How is testing accomplished?

Technical assistance
  1) What assistance is provided at the time of installation?
  2) Can staff training be conducted on-site?
  3) To what extent can the software and documentation be modified to meet user requirements?
  4) Who will make changes to the software and documentation?
  5) Will modification invalidate the warranty?
  6) Are any enhancements planned or in process?
  7) Will future enhancement be made available?

Quality practices
  1) Are the development and control processes followed?
  2) Are requirements, design, and code reviews used?
  3) If requirements, design, and code reviews are used, are they effective?
  4) Is a total quality program in place?
  5) If a total quality program is in place, is it documented?
  6) Does the quality program assure that the product meets specifications?
  7) Is a corrective-action process established to handle error corrections and technical questions?
  8) Is a configuration management process established?

Maintenance service
  1) Is there a guarantee in writing about the level and quality of maintenance services provided?
  2) Will ongoing updates and error conditions with appropriate documentation be supplied?
  3) Who will implement the updates and error corrections?
  4) How and where will the updates be implemented?
  5) What turnaround time can be expected for corrections?

Table C-7. *Continued*

Product usage
1) Can a demonstration of the software be made at the user site?
2) Are there restrictions on the purposes for which the product may be used?
3) What is the delay between order placement and product delivery?
4) Can documentation be obtained for examination?
5) How many versions of the software are there?
6) Are error corrections and enhancements release-dependent?

Product warranty
1) Is there an unconditional warranty period?
2) If not, is there a warranty?
3) Does successful execution of an agreed-upon acceptance test initiate the warranty period?
4) Does the warranty period provide for a specified level of software product performance for a given period at the premises where it is installed?
5) How long is the warranty period?

Costs
1) What pricing arrangements are available?
2) What are the license terms and renewal provisions?
3) What is included in the acquisition price or license fee?
4) What costs, if any, are associated with an unconditional warranty period?
5) What is the cost of maintenance after the warranty period?
6) What are the costs of modifications?
7) What is the cost of enhancement?
8) Are updates and error corrections provided at no cost?

Contracts
1) Is a standard contract used?
2) Can a contract be obtained for examination?
3) Are contract terms negotiable?
4) Are there royalty issues?
5) What objections, if any, are there to attaching a copy of these questions with responses to a contract?

b) Description of methods used to allocate system requirements, to develop the software requirements, architecture, and design, and to implement the source code and executable object code

c) Notations used to describe the system requirements and architecture

d) Notations used to describe the software requirements, architecture, design modules, design limitations, and code, including identification of the programming language(s) or subset used and reference to the definition of the language syntax, control, and data behavior, and side effects

e) Naming conventions for requirements, design, and source code, including source code, executable object code files, and data

Table C-8. Supplier performance standards [17]

Performance criteria
 1) Approach to meet software's functional requirements is defined.
 2) Growth potential or expansion requirement of the system is defined.
 3) Supplier meets time constraints for deliverables.
 4) Test and acceptance criteria that are to be met are defined.
 5) Programming language standards and practices to be followed are defined.
 6) Documentation standards to be followed are defined.
 7) Ease of modification is addressed.
 8) Maximum computer resources allowed, such as memory size and number of terminals, are defined.
 9) Throughput requirements are defined.

Evaluation and test
 1) Software possesses all the functional capabilities required.
 2) Software performs each functional capability as verified by the following method(s): documentation evaluation, demonstration, user survey, and test.
 3) Software errors revealed are documented.
 4) Software performs all system-level capabilities as verified by a system test.

Correction of discrepancies
 1) Supplier documents all identified discrepancies.
 2) Supplier establishes discrepancy correction and reporting.
 3) Supplier indicates warranty provisions for providing prompt and appropriate corrections.

Acceptance criteria
 1) All discrepancies are corrected.
 2) Prompt and appropriate corrections are provided.
 3) Satisfactory compliance to contract specifications is demonstrated by evaluations and test.
 4) Satisfactory compliance to contract specifications is demonstrated by field tests.
 5) All deliverable items are provided.
 6) Corrective procedures are established for correction of errors found after delivery.
 7) Satisfactory training is provided.
 8) Satisfactory assistance during initial installation(s) is provided.

Table C-9. Requirements traceability matrix example

CER #	Requirement name	Priority	Risk	Requirements document paragraph	Validation method(s)	Formal test paragraph	Status
1	Assignments and Terminations Module (A&T) Performance	2	M	3.1.1.1.1.1	Test Inspection Demonstration	4.1.2.6 4.1.2.7 4.1.2.9 4.1.2.3	Open

f) Methods of design and coding and constraints on design and code constructs and expressions, including design and code complexity restrictions and quality criteria for assessing requirements and design data and code

g) Presentation conventions and content standards for requirements data, design data, source code, and test data

h) Description of methods and tools used to develop safety monitoring software (if applicable)

i) Description of the methods and tools used to define traceability between system requirements, system architecture, software requirements, software architecture, design, code, and test elements

j) Description of methods, tools, and standards for testing

System Architectural Design Description

When a software product is part of a system, then the software life cycle should specify the system architectural design activity. The resulting system architectural description should identify items of hardware, software, and manual operations, and all system requirements allocated to the items. The following outline is provided by Section 6.25 of IEEE Std 12207.1, Software Life Cycle Processes Life Cycle Data [39].

Software Architectural Design Description

The resulting software architectural description describes the top-level structure and identifies the software components, which are allocated to all the requirements and further refined to facilitate detailed design.

The information provided here in support of design and development is designed to facilitate the definition of a software architectural description. This information was developed using IEEE 1471, Software Architectural Description, and addresses the ISO 9001 requirements that has been adapted to support ISO 9001 requirements. The modification of the recommended Software Architectural Description table of contents to support the goals of ISO 9001 more directly is shown in Table C-11. The

Table C-10. System architectural design document outline

a) Description
b) System overview and identification
c) Hardware item identification
d) Software item identification
e) Manual operations identification
f) Concept of execution
g) Rationale for allocation of hardware items, software items, and manual operations

Table 5-11. Software architectural design document outline

a) Generic description information
   Date of issue and status
   Scope
   Issuing organization
   References
   Context
   Notation for description
   Body
   Summary
   Glossary
   Change history
b) System overview and identification
c) Software item architectural design, including
   1) Software architecture general description
   2) Software component definition
   3) Identification of software requirements allocated to each software component
   4) Software component concept of execution
   5) Resource limitations and the strategy for managing each resource and its limitation
d) Rationale for software architecture and component definition decisions, including database and user interface design.

outline in the table is provided by Section 6.12 of IEEE Std 12207.1, Software Life Cycle Processes—Life Cycle Data [39].

Database Design Description

Many software efforts require the description of the database components. The outline in Table C-12 is provided by Section 6.4 of IEEE Std 12207.1, Software Life Cycle Processes—Life Cycle Data [39].

Software Architecture Design Success Factors and Pitfalls [64]

Through a series of software architecture workshops, Bredemeyer Consulting* has identified what they call the top critical success factors for an architecting effort. These are identified in Table C-13.

Conversely, the top pitfalls associated with an architecting effort have also been identified. These are shown in Table C-14.

*www.bredemeyer.com

Table C-12. Database design description document outline

a) Generic description information
    Date of issue and status
    Scope
    Issuing organization
    References
    Context
    Notation for description
    Body
    Summary
    Glossary
    Change history
b) Database overview and identification
c) Design of the database, including descriptions of applicable design levels (e.g., conceptual, internal, logical, physical)
d) Reference to design description of software used for database access or manipulation
e) Rationale for database design

Table C-13. Software architecting success factors

1. The strategic business objective(s) of the key sponsor must be addressed and the lead architect and team must have a sound understanding of functional requirements. There must be a good match between the technology and the business strategy.
2. The project must have a good lead architect with a well-defined role who:
    Exhibits good domain knowledge
    Is a good communicator/listener
    Is competent in project management
    Has a clear and compelling vision
    Is able to champion the cause
3. The architecting effort must contribute immediate value to the developers. The architecture must be based upon clear specifications.
4. There are architectural advocates at all levels of the organization, ensuring that there is strong management sponsorship and that required talent and resources are available.
5. Architecture is woven into the culture to the point that the validation of requirements occurs during each step of the process.
6. The customer is involved and expectations are clearly defined.
7. Strong interpersonal and team communication and ownership are present.

UML Modeling

Figure C-1 uses the UML notation to illustrate the inheritance, aggregation, and reference relationships between work product sets, their work products, their versions, the languages they are documented in, and the producers that produce them.

Unit Test Report

The unit test report template is designed to facilitate the definition of processes and procedures relating to unit test activities. This template was developed using IEEE Std 829, IEEE Standard for Software Test Documentation, and IEEE Std 1008, Software Unit Testing, which were adapted to support ISO 9001 requirements.

Table C-14. Software architecting pitfalls

1. The lead architect, or supporting senior management, does not effectively lead the effort. Roles and responsibilities are inadequately defined.
2. Thinking is at too low a level. The "big picture," or integration requirements are not considered.
3. There is poor communication inside/outside the architecture team.
4. There is inadequate support to include inadequate resources and/or talent.
5. There is a lack of customer focus. The architecture team does not understand or respond to functional requirements.
6. Requirements are unclear, not well defined, not signed off on, or changing.

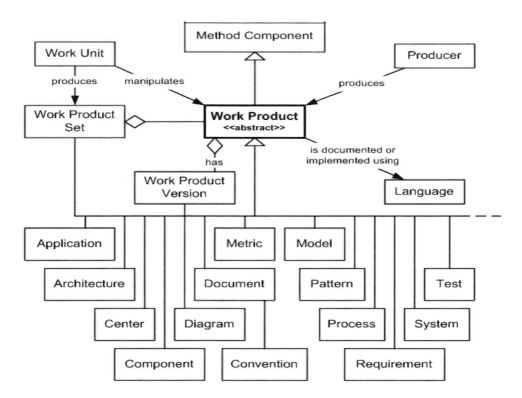

Figure C-1. UML notation.

Information displayed in italic text under each section is suggested items for inclusion in the unit test report. It is important to note that significant additional information is available from IEEE Std 829 and IEEE Std 1008, as the standards provide additional tutorial information. Table C-15 provides a sample document outline. The companion CD-ROM also provides an electronic template, *Unit Test Report.doc,* in support of this document.

Unit Test Report Document Guidance

The following provides section-by-section guidance in support of the creation of a unit test Report. This guidance should be used to help define a unit test process and should reflect the actual processes and procedures of the implementing organization. Additional information is provided in the document template, *Unit Test Report.doc,* that is located on the companion CD-ROM.

Table C-15. Unit test report document outline

1. Introduction
2. Identification and Purpose
3. Approach/Strategy
4. Overview
5. Test Configuration/Environment
   5.1   Software Environment
   5.2   Test Suite
   5.3   Test Stub
   5.4   Test Driver
   5.5   Test Harness
   5.6   Preservation of Testing Products
6. Test Cases/Items
   6.1   Test Case Identifier
   6.2   Test Sequence for Interdependencies
   6.3   Items Tested, Indicating their Version/Revision Level
   6.4   Specific Test Data
   6.5   Test Conditions Met
   6.6   Test Procedure/Script/Task Identifier Used
   6.7   Test Tool Used
   6.8   Results/Outputs/Summary of Faults
7. Evaluation
   7.1   Termination Requirements Met
   7.2   Test Effectiveness
   7.3   Variances from Requirement # and/or Design #
   7.4   Risks and Contingencies
8. Summary of Activities
9. Notes
10. Approvals

Introduction

This section should provide a brief introductory overview of the project and related testing activities described in this document.

Identification and Purpose

This section should provide information that uniquely identifies all test cases/items/component Description involved in the testing effort. This can also be provided in the form of a unique project plan unit test identifier. The following text provides an example:

> *The responsible tester was [Tester Name], [responsible organization], Role [coder, designer, test team member] and, unit test completed on [Date/time of test conclusion].*
>
> *This Unit Test Report # is to detail the unit testing conducted for the test items/component description in [Project Name] ([Project Abbreviation]) Version [xx], Statement of Work [date], Task Order [to number], Contract No. [Contract #], and Amendments.*
>
> *The goal of [project acronym] development is to [goal]. This development will allow [purpose].*

Approach/Strategy

This unit test is the first or bottom level of all tests planned according to the development life cycle. The unit test consists of functional testing, which is the testing of the application component(s) against a subset of the operational requirements. The unit test is classified as white-box testing, whose goal is to ensure that code is constructed properly and does not contain any hidden weaknesses.

This section should describe the unit test approach. Identify the risk areas tested. Specify constraints on characteristic determination, test design, or test implementation. Identify existing sources of input, output, and state data. Identify general techniques for data validation. Identify general techniques to be used for output recording, collection, reduction, and validation. Describe provisions for application software that directly interfaces with the units to be tested.

Overview

This section should provide a summary of all test items/components tested. The need for each item and its history may be addressed here as well. References to associated project documents should be cited here. The following provides an example:

> *[The list of individual program units (subprogram, object, package, or module) identifiers and their version/revision levels and configuration management Item(s) # are:*

Subprogram079   v15   Config12345
Subprogram081   v10   Config12346]

Specifics regarding the testing of these modules are identified in the:
    Any specific exclusions should be discussed.
    This document describes the Unit Test Report (UTR) for the [PRO-JECT ABBREVIATION] System.
    [PROJECT ABBREVIATION] documentation related to this UTR are:

- *Project Plan Milestone/Task #*
- *Software Requirements Traceability Matrix (RTM) line item descriptions:*
    - *Software Requirements Specification (SRS)*
    - *Software Development Plan (SDP)*
    - *Software Design Document (SDD)*
- *User's Manual/System Administrator's Manual,*
- *Data Dictionary*
- *Test Problem Report (TPR) #*
- *Change Enhancement Request (CER) #*

This UTR describes the deliverables to be generated in the testing of the [PROJECT ABBREVIATION]. This report complies with the procedures defined in the [PROJECT ABBREVIATION] Software Configuration Management (SCM) Plan and the [PROJECT ABBREVIATION] Software Quality Assurance (SQA) Plan. Any nonconformance to these plans will be documented as such in this UTR.

This document is based on the [PROJECT ABBREVIATION] Unit Test Report template with tailoring appropriate to the processes associated with the creation of the [PROJECT ABBREVIATION]. The information contained in this UTR has been created for [Customer Name] and is to be considered "For Official Use Only."

Test Configuration/Environment

Provide a high-level summary of test configuration/environmental information. Depending on system complexity, this could include:

- Software Environment
- Test Suite
- Test Stub
- Test Driver
- Test Harness
- Preservation of Testing Products

Software Environment

This subsection should provide a complete list of all communications and system software, Operating system version, physical characteristics of the facilities, including the hardware, and database version. System and software items used in support of testing should include versions. If the system is kept in an online repository, reference to the storage location may be cited instead of listing all items here. The following are examples:

> *System and software used in the testing of [PROJECT ABBREVIATION]: [System and software List].*
>
> *For details of client and server system specifications see the [PROJECT ABBREVIATION] System Requirements Specifications (SysRS).*
>
> *Unit tests were to be performed at [company name], [site location] in the test system [test system number], using OS X, 10.3.2.*

Test Suite

Describe the set of test cases that serve a particular testing goal.

Test Stub

Describe the temporary, minimal implementation of a component to increase controllability and observability in testing.

Test Driver

Describe the class or utility program that applies test cases to the component.

Test Harness

Describe the system of test drivers and other tools to support test execution.

Preservation of Testing Products

Describe the location of resulting test data, any revised test specifications, and any revised test data.

Test Cases/Items

This section should describe each of the test cases/items. Depending on system complexity, this could include:

- Test case identifier
- Test sequence for interdependencies

- Items tested, indicating their version/revision level
- Specific test data
- Test conditions met
- Test procedure/script/task identifier used
- Test tool used
- Results/outputs summary of faults

Test case identifier

This subsection should describe the unique identifier of the test case, allowing others to find and repeat the same test case, for example, Test Case 051.

Test Sequence for Interdependencies

This subsection should describe where the sequence of this test case fits in the hierarchy or architecture of other test cases identifiers. The following is provided as an example:

Test Case 052 was the 2nd of 8 test cases following Test Cases 04x and preceding Test Cases 06x.

Items Tested, Indicating their Version/Revision Level

This subsection should describe the test items/components tested by this test case. The following is provided as an example:

Subprogram079   v15   Config12345
Subprogram081   v10   Config12346]

Specific Test Data

This subsection should describe the input data and states used by this test case (e.g., queries, transactions, batch files, references, software data structures, or invalid inputs). The following is provided as an example:

Subprogram081   testfile123456

Test Conditions Met

This subsection should describe the resulting test condition for the test case; the test condition compares the values of two formulae. If the comparison yields true, then the condition is true. Otherwise, it is false. Some typical conditions may be:

- Check for correct handling of erroneous inputs
- Check for maximum capacity
- User interaction behavior consistency
- Retrieving data
- Saving data
- Display screen and printing format consistency
- Measure time of reaction to user input

The following is provided as an example:

Subprogram081   *True, created 256 byte testfile123457*

Test Procedure/Script/Task Identifier Used

This subsection should describe the tasks conducted for this test case. Also note if there are any constraints on the test procedures, or special procedural requirements, or unusual resources used. The following is provided as an example:

Task062   *initiate Subprogram079, followed by Subprogram081*

Test Tool Used

This subsection should describe the test case tool use from the test environment. The following is provided as an example:

Test Suite Alpha, Test Driver A1 of Test Harness A

Results/Outputs/Summary of Faults

This subsection should describe the results of using this test case. For each failure, have the failure analyzed and record the fault information:

- Case 1: Test Specification or Test Data
- Case 2: Test Procedure Execution
- Case 3: Test Environment
- Case 4: Unit Implementation
- Case 5: Unit Design

The following is provided as an example:

Task062   *revised to v8 (Case 2)*
Testfile123456   *revised to v3 (Case 1)*
Subprogram081   *revised to v8, v9, and v10 (Case 4)*

Evaluation

Termination Requirements Met. This subsection should provide a list of the termination requirements met by the unit testing, starting with a summary the evaluation of the test cases/items.

Test Effectiveness. This subsection should Identify the areas (for example, features, procedures, states, functions, data characteristics, and instructions) covered by the unit test set and the degree of coverage required for each area. Coverage can be stated as a percentage (1 to 100%) for the major objective units:

- Code
- Instructions
- Branches
- Path
- Predicate
- Boundary value
- Features
- Procedures
- States
- Functions

Variances from Requirement # and/or Design #. This subsection should report any variances of the test items from their design and/or requirements specifications. If this occurred, specify the:

- Reason for each variance
- Identification of defects that are not efficiently identified during previous peer reviews
- Resulting test problem report (TPR) #s

Risks and Contingencies. This subsection should identify any unresolved incidents and summarize the test item limitations.

Summary of Activities

This section should provide actual measurements for calculation of both test efficiency and test productivity ratios:

- Hours to complete testing
  - Not counting test suspension
- Hours for setup
- Hours test environment used
- Size of testing scope

Notes

This section should provide any additional information not previously covered in the unit test report.

Approvals

This section should include the approvals specified in the project plan.

System Integration Test Report

This system integration test report template is designed to facilitate the definition of processes and procedures relating to system integration test activities. This template was developed using IEEE Std 829-1998, IEEE Standard for Software Test Documentation, and IEEE Std 1008-1987, Software Unit Testing, which were adapted to support ISO 9001 requirements.

Information displayed in italic text under each section is suggested items for inclusion in the system integration test report. It is important to note that significant additional information is available from IEEE Std 829-1998 and IEEE Std 1008-1987, as the standards provide additional tutorial information. Table C-16 provides a sample document outline. An electronic version of this document is provided on the companion CD-ROM, identified by *System Integration Test Report.doc.*

System Integration Test Report Document Guidance

Introduction

This section should provide a brief introductory overview of the project and related testing activities described in this document.

Table C-16. System integration test
report document outline

1. Introduction
2. Identification and Purpose
3. Approach/Strategy
4. Overview
5. Test Configuration
    5.1 Test Environment
    5.2 Test Identification
    5.3 Results/Outputs/Summary of Faults
6. Test Schedules
7. Risk Management/Requirements Traceability
8. Notes
9. Approvals

Identification and Purpose

This section should provide information that uniquely identifies the system and the associated testing effort. This can also be provided in the form of a unique project plan milestone identifier. The following text provides an example:

> *The responsible lead tester was [Tester Name], of [Responsible organiza-tion], with role [lead test team member title]. System Integration Test was completed on [Date/time of System Integration Test conclusion], as planned in the System Integration Test Plan.*
>
> *This System Integration Test Report is to detail the System Integration testing conducted for the [Project Name] ([Project Abbreviation]) Ver-sion [xx], Statement of Work [date], Task Order [to number], Contract No. [Contract #], and amendments.*
>
> *The goal of [project acronym] development is to [goal]. This system will allow [purpose].*
>
> *Specifics regarding the implementation of these system modules are identified in the [Project Abbreviation] System Requirements Specifica-tion (SysRS) with line item descriptions in the accompanying Require-ments Traceability Matrix (RTM) and references to the Interface Control Documentation (ICD) and the System Integration Test Plan (SysITP).*

Approach/Strategy

This system integration test follows unit testing of all modules/components sched-uled for integration, as planned according to the system integration test plan. The re-sult is the creation of either a partial or complete system. These tests result in discov-ery of problems that arise from modules/components interactions and these problems are localized to specific components for correction and retesting. The system inte-gration test includes white-box testing, whose goal is to ensure that modules/compo-nents are constructed properly and do not contain any hidden weaknesses.

This section should summarize the deviations from the system integration test plan objectives, kinds (software, databases, hardware, COTS, and prototype usabil-ity), and approach (e.g., top-down, bottom-up, object-oriented, and/or interface test-ing).

Overview

This section should provide a summary of all system modules/components and sys-tem features tested. Any specific exclusion should be discussed. The following pro-vides an example:

> *The list of individual program units (subprogram, object, package, or module) identifiers and their version/revision levels and configuration management item(s) # are:*

Subprogram078   *v19*   *Config12343*
Subprogram079   *v15*   *Config12345*
Subprogram080   *v13*   *Config12344*
Subprogram081   *v10*   *Config12346*

This document describes the System Integration Test Report (SysITR) for the [PROJECT ABBREVIATION] System. [PROJECT ABBREVIATION] documentation related to this SysITR is:

Project Plan Milestone/Task #
System Requirements Specification (SRS),
System Requirements Traceability Matrix (RTM)
Test Problem Report (TPR) #
Change Enhancement Request (CER) #

This SysITR describes the deliverables to be generated in the system integration testing of the [PROJECT ABBREVIATION]. This report complies with the procedures defined in the [PROJECT ABBREVIATION] Software Configuration Management (SCM) Plan and the [PROJECT ABBREVIATION] Software Quality Assurance (SQA) Plan. Any nonconformance to these plans will be documented as such in this SysITR.

This document is based on the [PROJECT ABBREVIATION] System Integration Test Report template, with tailoring appropriate to the processes associated with the creation of the [PROJECT ABBREVIATION]. The information contained in this SysITR has been created for [Customer Name] and is to be considered "For Official Use Only."

Test Configuration

This section should summarize the deviations from the System Integration Test Plan—System Integration Test—Environmental. The following are examples:

The System Integration Tests were performed at [company name], [site location]. All integration testing were conducted in the development center [room number]. The deviations from the System Integration Test Plan for the system and software used in the testing of [PROJECT ABBREVIATION] are asterisked below:

[System and software List]

Several errors in Test Suite Alpha, Test Driver A1 of Test Harness A were discovered and corrected. Resulting retesting at both unit and subsystem levels added one day to the test interval, but no additional risk.

Test Identification

This subsection should summarize the deviations from the System Integration Test Plan—Test Identification. The following is an example:

Subsystem A integration was delayed due to late arrival of Module A12345. This resulted in resequencing Subsystem B integration testing. A one week longer test interval resulted, but no additional risk.

Results/Outputs/Summary of Faults

This subsection should describe the results of system integration testing. The following are examples:

The 35 failures analyzed and localized to specific components or the test environment resulted in the following fault information:

- *Case 1: Test Specification or Test Data*
- *Case 2: Test Procedure Execution*
- *Case 3: Test Environment*
- *Case 4: Component Implementation*
- *Case 5: Component Design*

The above seven components with defects and corrected version #s were:

Subprogram078	*v19*	*Config12343*
Subprogramxxx	*vxx*	*Configxxxxx*
. . .		

According to the System Integration Test Plan measurement objectives, 85% of all test scripts passed on the first pass.

Test Schedules

This section should summarize the deviations from the System Integration Test Plan—Test Schedule.

Risk Management/Requirements Traceability

This section should identify all high-risk items encountered and the action taken. The following provides an example:

While all components have been tested and integrated, with only two weeks added to the test interval, several changes are needed in the System Test Plan to fully complete the revised Requirements Traceability Matrix.

Summary of Activities

This section should provide actual measurements for calculation of both test efficiency and test productivity ratios:

- Hours to complete testing
  - ○ Not counting test suspension
- Hours for setup
- Hours test environment used
- Size of testing scope

Notes

This section should provide any additional information not previously covered in the system integration test report.

Approvals

This section should include the approvals specified in the project plan.

OPERATION

Product Packaging Information

The following example of a product description is based upon the recommendations of IEEE Std 1465, IEEE Standard Adoption of ISO/IEC 12119—Information Technology—Software Packages—Quality Requirements and Testing [32]. Table C-17 provides an example of a product description and describes the types of information that should be present in every product description.

MAINTENANCE

Change Enhancement Requests

The identification and tracking of application requirements is accomplished through some type of change enhancement request (CER). A CER is used to control changes to all documents and software under SCM control and for documents and software that have been released to SCM. Ideally, the CER is an online form that the software developers use to submit changes to software and documents to the SCM Lead for updating the product. Table C-18 provides an example of the types of data that may be captured and tracked in support of application development or modification.

Baseline Change Request

The tracking of changes to the development baseline may be accomplished using a baseline change request (BCR). A BCR is a variation of the basic CER that is used to control changes to all documents and software under SCM control and for docu-

Table C-17. Product description

Product description sheet, SecurTracker Version 1.0, an automated security tool.

What is SecurTracker?

SecurTracker is a *comprehensive* security management tool that incorporates the efforts of security professionals, administrative staff, IS team, and management team to integrate a complete security solution.

Architectural Goals

SecurTracker will support five architectural goals, identified below:

1. Meet user requirements
2. Promote system longevity
3. Support incremental system implementation
4. Aid iterative system refinement
5. Provide mechanisms for system support and maintenance

Why Use SecurTracker?

SecurTracker is the only application available that supports collateral, SCI, and SAR environments.

SecurTracker is a multi-user application that maintains all security management data in one centralized location. Centralization offers numerous significant advantages, including the elimination of duplicate records storage, reduction of risk related to inconsistent data, and a decrease in the amount of storage space (hardware costs) required to maintain security records.

SecurTracker was written using security professional expertise. Software development teams who read security operating manuals to determine software requirements write other competitive security management tools. This approach inevitably omits the dynamic, real-world situations that cannot be captured in an operating manual.

SecurTracker is an interactive Web-based solution. Client setup and maintenance is not required. Setup, configuration, and updates are accomplished at the server level, thus eliminating costly and time-consuming system maintenance.

SecurTracker Implementation Approach

SecurTracker is available as a single-user, stand-alone system or as a multiuser, Web-based product. The user can accomplish the stand-alone installation. The installation process for multiuser implementation offers a two-phase approach; the first phase would consist of the database and Web server setup; the second phase would include data migration, validation, and testing.

The implementation team will begin by extracting the security data from the legacy system(s). Once this is accomplished, the team will then extract and incorporate into SecurTracker the data. Data extraction can be accomplished remotely, thus eliminating travel cost.

After data extraction is accomplished, the SecurTracker system will be tested to verify the accuracy of the data conversion. Any necessary adjustment to the data import functions will be made at this time. Once data has been successfully extracted from one legacy system, the extraction process can be automated for other like systems, creating an efficient, repeatable process.

(continued)

Table C-17. *Continued*

During the implementation phase, the implementation team will work closely with the on-site team lead and functional experts to determine an approach that ensures data integrity during migration. Final deployment includes user setup and maintenance item customization.

Data conversion, user training, and phone support are available on all solutions!

Points of Contact

Susan K. Land
SecurTracker Project Manager
(256) xxx—xxxx
e-mail address

System Requirements

Database Server*
1 GB RAM; 60 GB Hard Drive; 1 GHz processor; Network Card, Windows NT 4.0 (SP6a)/Windows 2000 Server (SP3); Microsoft Excel; Oracle 8.1.6 or Oracle 9i.

Web Server*
256 MB RAM; 20 GB Hard Drive; 700 MHz processor; Network Card; Windows NT 4.0 (SP6a)/Windows 2000 Server (SP3 and security patches); Oracle Forms and Reports 6i; Outlook Express Version 5.0/Microsoft Outlook 4.0 (with latest SP).

Web Client Machines
A Web Client machine can be any Internet-connected system that utilizes a browser; Video card supporting 256 colors, 800 × 600 resolution (or greater); Network Card.

*Although it is highly recommended that the Web and Database servers run on separate machines, a Web/Database server can operate on one machine. Using this option, the hardware requirements listed for the database server must be in place. Additionally, ALL software listed for the Database AND Web Server specifications must be installed on this server.

Standalone Configuration
512 MB RAM; 20 GB Hard Drive; 700 MHz processor; Windows NT 4.0 (SP6a/Windows 2000 (SP3); Personal Oracle 8.1.6 or 9i; Outlook Express 5.0/Microsoft Outlook 4.0; IE 5.5.

Frequently Asked Questions

Q: What languages/technologies were used in development?
A: SecurTracker is currently written in Oracle Forms and Reports 6i using an Oracle 9i database.

Q: Can the database be migrated to other versions of Oracle?
A: Yes. SecurTracker supports Oracle Standard and Enterprise editions of Oracle 8.1.6 or 9i.

Q: How is reporting implemented?
A: Reporting to Web, printer, e-mail are implemented by Oracle Reports and leverages .PDF format.

Q: What is the hard drive footprint for the Web server?
A: 400 MB, which includes the application code.

Table C-17. *Continued*

Q: What is the hard drive footprint for the database server?
A: This depends upon the amount of space allocated for data. Initial setup is estimated at 500 MB.

Q: How is the standalone version different from the Web version?
A: The standalone version will use Personal Oracle, which supports a single user. The Web version uses Oracle 8i or 9i, which will support any number of licensed users.

Q: Is there a limit on the number of records the system can handle?
A: The only limits occur by the size of the hard drive; therefore, by adding more physical space as needed, the number of records is limitless.

ments and software that have been released to SCM. Ideally, the BCR is an online form that the software developers use to submit changes to software and documents to the SCM Lead for updating the product. Table C-19 provides an example of the types of data that may be captured and tracked in support of application development or modification. *Baseline Change Request.doc* provides an electronic version of this BCR on the companion CD-ROM.

Work Breakdown Structure for Postdeployment

The example provided in Table C-20 provides suggested content for a WBS in support of the postdevelopment stage of a software development life cycle. This WBS content is illustrative only and should be customized to meet specific project requirements.

Configuration Management

Configuration Control Board (CCB) Letter of Authorization. The following is provided as an example of a formal letter of CCB authorization. This letter may be used to help officially sanction CCB groups. An electronic version of this letter is provided on the companion CD-ROM, entitled *CCB Letter of Authorization.doc.*

MEMORANDUM FOR [Approving Authority]

DATE: [Date]

FROM: [Company Name]
       [Program Name]
       [Name, Program Manager]
       [Name, Project Manager

SUBJECT: Establishment of [Project Name] Configuration Control Board (CCB)

1. This is to formally propose the charter of the [Project Name] Configuration Control Board (CCB).

Table C-18. Typical elements in a CER

Element	Values
CER #	Unique CER identifier assigned by CRC.
Type	One of the following:
	BCR—Baseline Change Request; CER indicating additional/changed requirement.
	SPR—Software Problem Report; CER indicating software problem identified externally during or post Beta test.
	ITR—Internal Test Report; CER indicating software problem identified internally through validation procedures.
	DOC—Documentation Change; CER indicting change to software documentation.
Status	One of the following:
	OPEN—CER in queue for work assignment.
	TESTING—CER in validation test.
	WORKING—CER assignment for implementation as SCR(s)
	VOIDED—CER deemed not appropriate for software release.
	HOLD—CER status to be determined.
	FIXED—CER incorporation into software complete.
Category	One of the following:
	DATA—Problem resulting from inaccurate data processing.
	DOC—Documentation inaccurate.
	REQT—Problem caused by inaccurate requirement
Priority	One of the following:
	1—Highest priority; indicates software crash with no work-around.
	2—Cannot perform required functionality; available work-around.
	3—Lowest priority; not critical to software performance.
	0—CER has no established priority.
Date Submitted	The date CER initiated; defaults with current date.
Date Closed	The date CER passes module test.
Originator	Identifies the source of CER.
Problem Description	Complete and detailed description of the problem, enhancement, or requirement.
Short Title	Short title summarizing CER.
SRS Ref#	Cross-reference to associated software requirements specification (SRS) paragraph identifier.
SDD Ref#	Cross-reference to associated software design document (SDD) paragraph identifier(s).
STP Ref #	Cross-reference to associated software test plan (STP) paragraph identifier(s).
Targeted Version	Software module descriptor and version number; determined by SCM Lead.

Table C-18. *Continued*

Element	Values
File	Files affected by CER.
Functions Affected	List of functions affected by CER.
Time Estimated	Estimated hours until implementation complete.
Time Actual	Actual hours to complete CER implementation.
Module	Name of module affected by CER.
Prob. Description	Description of original CER; add comments regarding implementation.
Release	Release(s) affected by CER.
CER#	Cross-reference number used to associate item with other relevant CERs.
Software Engineering Name	Last name of engineer implementing CER.

2. It is our desire that the [Project Name] CCB be chartered (Attach.1) with the authority to act as the management organization responsible for ensuring the documentation and control of [Project Name] software development by ensuring proper establishment, documentation and tracking of system requirements. In addition, the CCB will serve as the forum to coordinate software changes between represented agencies and assess the impacts caused by requirements changes.

3. Those recommended to participate in CCB voting will be [Approving Authority] housing representatives from the following:

Management Board:
[Organization Name]

Beta Test Site [Organization Name] Representatives:
[List of sites]

Contractor Associates:
[Company Name]
[Subcontractor Name]

4. The CCB will be the necessary "single voice," providing clear direction during [Project Name] requirements definition, product design and system development. It is our goal to produce the highest-quality product for the [Project Name] user. It is our desire that the [Project Name] CCB be chartered to provide [Project Name] development with consolidated and coordinated program direction.

Configuration Control Board Charter. The following is provided as the suggested example content and format of a configuration control board (CCB) charter. This charter may be used to establish project oversight, control, and

Table C-19. Baseline change request

Originator	Module:		Priority:
Name:	Title:		BCR No.:
Organization:	SRS Revision:		Date:

BCR Title:

Change Summary:

Reason For Change:

Impact if Change Not Made:

Org	Name	Yes	No	Desired Applicability
				Release Available:
				Release Unsupported:
				CCB Approval
				Name:
				Date:
				Signature:
				_____ Approved
				_____ Disapproved
				Comments:

Table C-20. WBS content for postdevelopment activity

Product Installation
  Distribute Product (software)
    Package and distribute product
    Distribute installation information
    Conduct integration test for product
    Conduct regression test for product
    Conduct user acceptance test for software
    Conduct reviews for product
    Perform configuration control for product
    Implement documentation for product
  Install Product (software)
    Install product (packaged software) and database data
    Install any related hardware for the product
    Document installation problems
  Accept Product (software) in Operations Environment
    Compare installed product (software) to acceptance criteria
    Conduct reviews for installed product (software)
    Perform configuration control for installed product (software)
Operations and Support
    Utilize installed software system
    Monitor performance
    Identify anomalies
    Produce operations log
    Conduct reviews for operations logs
    Perform configuration control for operations logs
Provide Technical Assistance and Consulting
  Provide response to technical questions or problems
  Log problems
Maintain Support Request Logs
  Record support requests
  Record anomalies
  Conduct reviews for support request logs
Maintenance
  Identify Product (software) Improvements Needs
    Identify product improvements
    Develop corrective/perfective strategies
    Produce product (software) improvement recommendations
  Implement Problem Reporting Method
    Analyze reported problems
    Produce report log
    Produce enhancement problem reported information
    Produce corrective problem reported information
    Perform configuration control for reported information
  Reapply Software Life Cycle methodology

responsibility. An electronic version of this charter example is provided on the companion CD-ROM, entitled *CCB Charter.doc*.

Scope. This Configuration Control Board (CCB) Charter establishes guidelines for those participating in the (Project Name) CCB process. Any questions regarding this document may be directed to (Chartering Organization/Customer).

CCB Overview. The (Project Name) CCB is the management organization responsible for ensuring the documentation and control of (Project Name) software development by ensuring proper establishment, documentation, and tracking of system requirements. In addition, the CCB serves as the forum to coordinate software changes between represented agencies and assess the impacts caused by these changes.

CCB Meetings. The (Project Name) CCB will be conducted quarterly and will include the following agenda items:

a. Recommended changes to the CCB Charter
b. (Project Name) schedule
c. Discussion of proposed Baseline Change Requests (BCRs) for approval/disapproval
d. Review of new BCRs
e. Status of open action items
f. Review of new action items
g. Program risk assessment

Meeting dates and times will be determined by CCB consensus and announced at least 60 days prior to any scheduled meeting. Reminders will be distributed via DoD Messages, electronic mail, memoranda, and/or bulletin board announcements.

Formal minutes will be available and distributed within 30 days of the meeting's conclusion.

A CCB report will be prepared and maintained by the CCB chairman, listing CCB membership; individual representatives; interfacing modules, subsystems, or segments; pending or approved BCRs; status of software development activities such as open action items and risk tracking; and any other identified pertinent (Project Name) program information. Table C-21 provides an example of a BCR.

Participation. CCB participation is open to all government and contractor organizations that develop or use (Project Name) associated systems, or (Project Name) elements. Participants belong to one of six categories:

1. Management Board (MB)
2. Chairman
3. Board Members (Command Representatives)
4. Implementing Members
5. Associate Members
6. Advisors

Table C-21. Example Baseline Change Request (BCR)

Originator	Module:	
Name:	Title:	BCR No.:
Organization:	SRS Revision:	Date:

BCR Title:

Change Summary:

Reason For Change:

Impact if Change Not Made:

Org	Name	Yes	No	Desired Applicability
				Release Available:
				Release Unsupported:
				CCB Approval
				Name:
				Date:
				Signature:
				_____ Approved
				_____ Disapproved
				Comments:

Discussions concerning (Project Name) BCRs will be limited to the organizations responsible for the implementation of the BCRs. If other organizations are interested in the operational usage of the BCRs, these issues should be brought to the attention of the Board Members.

Participants. The MB is composed of the following participants. (*This should be three or less people at a very high level with the ability to make command decisions for the CCB.*)

The purpose of the MB is to address software development problems that cannot be resolved at the CCB Board level. The MB will be chaired by _____ and co-chaired by _____.

Chairman _____ is responsible for establishing and running the CCBs in accordance with this plan that includes the following:

a. Leading the effort to document and control all module items associated with HIMS.
b. Ensuring a proper forum exists for coordinating BCRs and selecting the solution that is in the best interest of the Government.
c. Ensuring minutes and action items are recorded and distributed.

The Chairman has the final approval authority for all BCRs.
Board Members: Government organizations are responsible for the following:

a. Reviewing BCRs to ensure that they support operational requirements.
b. Voting to approve/disapprove BCRs.

Implementing Members: Government organizations on contract are responsible for the following:

a. Implementing BCRs.
b. Reviewing BCRs for adequacy and program impact (cost or schedule).

Associate Members: Government organizations not on contract are responsible for BCRs for adequacy.

Advisors: Government or contractor agencies invited to a specific CCB to support discussion of a specific BCR are responsible for presenting information relevant to a BCR when requested by the sponsoring Operational, Implementing, or Associate Member.

CCB Process

General: All CCB actions are coordinated and take place through the BCR process. Any CCB Participant may propose BCRs, but proprietary BCRs will not be accepted or discussed. It is important to understand that (Project Name) technical interchange meetings (TIMs) provide the forum to brainstorm and discuss preliminary BCR ideas prior to the formal process described in this document. The CCB acts only as the final review and adjudication for BCRs.

<u>Software Requirements Specifications (SRSs):</u> (Project Name) software requirements are brought under CCB control in the BCR process. A BCR describing the request for a software requirements baseline is generated and distributed at the CCB. The BCR is reviewed using the accepted practices and, when approved, the SRS is accepted for CCB management.

<u>BCR Format:</u> BCRs are prepared by generating "is" pages and attaching those to a BCR cover sheet to the document. "Is" pages are change pages indicating how the page should read. A BCR request form is provided at the end of this document for reference.

<u>BCR Process:</u> BCRs can be coordinated using any of the following methods:

  a. BCRs will be provided to _____ at least thirty (30) days prior to a CCB meeting so they may be reproduced and included in the agenda. Please fax to: (*the BCRs should be sent to the Chair and distributed to all participants*).
  b. The Board, Implementing, or Associate Member sponsoring the BCR will be provided the opportunity to present a five (5) minute overview of why the change is needed and what is being changed.
  c. Any Board, Implementing, or Associate Member that is potentially impacted by the BCR will identify himself/herself as a reviewing member for the BCR.
  d. Reviewing members will submit their concurrence or nonconcurrence to the sponsoring member and Chair. (Silence is considered concurrence.) If a reviewing member nonconcurs, any issues associated with implementing the BCR, including alternate solutions, will be documented and sent to the sponsoring member.
  e. If issues are identified, the sponsoring member will work with all reviewing members to try to resolve conflicts prior to discussion at the CCB.
  f. At the CCB, each potential BCR is reviewed for concurrence/nonconcurrence. BCRs that all reviewing agencies concur upon will be signed by the Board members and approved by the Chairman. These BCRs will be prioritized and scheduled for implementation.
  g. BCRs that have received nonconcurrence go through a formal review. The formal review consists of each reviewing member presenting a ten (10) minute presentation explaining their concurrence or nonconcurrence with the BCR, including their preferred approach and, contractual/cost/schedule impacts.
  h. The Chairman will decide whether to approve, disapprove, or defer the BCR. If the BCR is deferred, the process is repeated again starting with "d" above and the reviewing agencies will use this time to attempt to reach concurrence.

<u>Accelerated Process:</u> (Emergency BCR)

  a. BCRs will be provided to _____ and all affected CCB members at least forty-five (45) days prior to the CCB. Reviewing members will submit their concurrence or non-concurrence to the sponsoring member and Chair. (Silence is considered concurrence.)

b. At the CCB, each potential BCR is reviewed for concurrence/nonconcurrence. BCRs that all reviewing agencies concur upon will be signed by Implementing Members and approved by the Chairman.

c. BCRs that have received nonconcurrence will go through a formal review. The formal review consists of each reviewing member presenting a ten (10) minute presentation explaining why they concur or nonconcur with the BCR, including their preferred approach and contractual/cost/schedule impacts.

d. The Chairman will decide whether to approve, disapprove, or defer the BCR. If the BCR is deferred, the process is repeated starting with "c" above. The reviewing agencies will use this time to attempt to reach concurrence.

Contractual Direction. After a decision has been made by the CCB, all Implementing Members affected by the change will take the contractual action needed to ensure that the change is implemented.

Software Change Request Procedures

A change to software may be requested by one of the following change enhancement requests (CER): internal test report (ITR), software problem report (SPR), or baseline change request (BCR) from external or internal sources or from the product specification/requirements. A document change request (DCR) may be used to request a change to documentation. All of the above requested change procedures are referred to as a change request in the text below. These procedures are outlined in Figure C-2 for [Project Abbreviation] maintenance and production life cycles.

The Configuration Control Board (CCB) for each development team determines which of the change requests is required for each software release prior to the start of work for that release. Additional change requests are reviewed by [Project Abbreviation] CCB to determine and assign the proper status to the change requests, these are held for CCB scheduling. Status is one of the following:

Open—Change request is to be implemented for current software version.

Hold—Change request targeted for another software version.

Voided—Change request is a duplicate of an existing CER or does not apply to existing software.

Working—Change is currently being implemented.

Testing—Implemented change request is under DT&E evaluation.

Fixed—Change request is implemented, unit and integration test complete.

The following steps define the procedure for each status. Each procedure starts with the change request being given to the Project Lead or designated Change Request Coordinator (CRC).

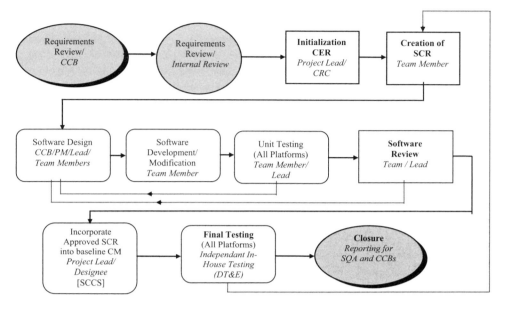

Figure C-2. Software baseline change process.

A. Open
   1. Initialization
      a) The CRC for the associated project creates a CER.
      b) The CRC notifies the Project Lead of the new CER.
   2. Create SCR
      a) The PL reviews the CER, and provides estimates of schedule impact to CCB.
      b) CCB prioritizes CER and authorizes implementation.
      c) The PL assigns the CER to a Software Engineer (SE) to incorporate.
      d) The SE creates a Software Change Request (SCR).
   3. Software Design
      The SE designs changes necessary to implement CER:
      a) The SE identifies interface changes and consults with appropriate PLs.
      b) Determines changes needed.
      c) The SE identifies any changes needed for documentation.
      d) The SE estimates work effort required for CER completion.
   4. Software Development/Modification
      a) The SE makes needed changes and provides updated documentation for testing.
      b) The SE checks out the files from the configuration managed version (not the baselined version) of all the files needed to incorporate the CER.
      c) The SE fills in the CM sections of the SCR.
      d) The SE puts the changes in the configuration managed files.

5. Unit Testing
   The SE tests on all platforms, if test fails, go to Step 3, Software Design.
6. Software Review
   a) The SE prepares a software review package:
      1) Hard copy of SCR
      2) List of documentation changes
   b) Software Review
      1) SE gives software review package to another team member for review.
      2) If SCR peer approved, go to Step 7.
      3) SCR not approved, go to Step 3, Software Design.
7. Incorporate Approved SCR into Baseline CM
   a) The PL changes the SCR status to indicate that it is approved.
   b) The SE checks in the files.
   c) The SE fills in the CM sections of the SCR.
   d) The CRC reviews SCR for completeness.
   e) The CRC changes the status on the CER to indicate testing.
   f) The designated SCM gives a list of changed files to PLs for updating their working directories/executables.
   g) The designated SCM installs changes to baseline on all platforms.
8. Final Testing
   a) The CRC identifies all CERs with testing status.
   b) Final testing by DT&E validation group (If problems are found, start over at Step 2a).
9. Closure
   The CRC changes final status on CER to indicate complete.
B. Hold
   The CRC creates a CER marking status HOLD and indicates the targeted version for incorporating the change. The CRC files the original change request for the next version. The change will be considered for inclusion in the next version.
C. Voided
   1. Duplicates:
      a) The CRC marks the change requests as a duplicate, specifying which CER is a duplicated.
      b) The CRC files the change request for associated version.
   2. Other:
      a) The CRC marks the reason for cancellation.
      b) The CRC files the change request for associated version.
D. Testing
   The CRC forwards CER/SCR to PL for review. After review the PL forwards CER to DT&E test for validation. CER status is updated to indicate testing status.
E. Fixed
   The CRC files the closed change request and updates database. Figure C-3 provides an overview of the software change request procedures.

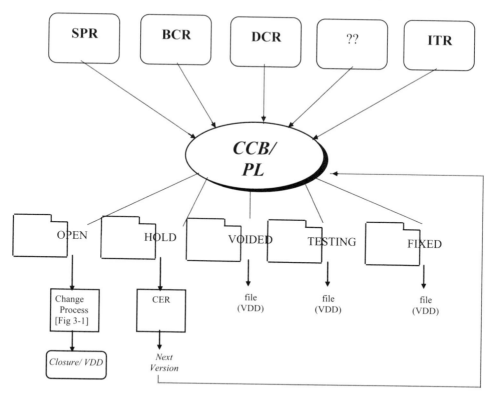

Figure C-3. Software change request procedures.

QUALITY ASSURANCE

Example Life Cycle

The following describes a full development life cycle for a new [project abbreviation] software product and recommended software quality assurance activities (see Figure C-4). [Note: This information may be described in this SQA plan, or in the associated SPMP.] See Table C-22 for a tabular summary of responsible parties:

Step (0)

Tasking is the process of receiving a statement of work (SOW) from the customer. It is the official requirement to which all software development efforts must be directed. The customer, [customer name] receives requirements from its associated Configuration Control Board (CCB).

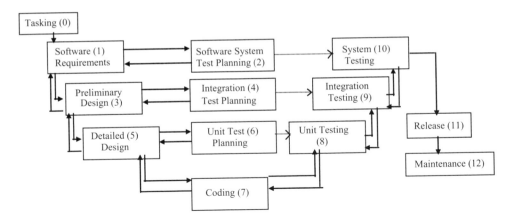

Figure C-4. Development life cycle.

Table C-22. Software development life cycle—responsibility

Life cycle activity	Review by	Product
(0) Tasking	CCB	Work Request
		ICWG Minutes
(1) Software Requirements	[Customer Name], PM, PL, CCB	SRS
		SRS Review
(2) Software System Test Planning	[Customer Name], PM, PL, CCB	Software System Test Plan
		Software System Test Plan Review
Software Project Plan	[Customer Name], PM, PL, CCB	Software Project Plan
(3) Preliminary Design	[Customer Name], PM, PL, CCB	Draft Design Document
		Draft Design Document Review
(4) Integration Test Planning	[Customer Name], PM, PL, CCB	Software Test Plan
		Software Test Plan Review
(5) Detailed Design	[Customer Name], PM, PL, CCB	Software Project Plan
		SPP Review
(7) Coding	PL, SQAM	Audit Documentation
	PL, SQAM	Walk-through Documentation
(8–10) Unit, Integration, System Testing	[Customer Name], PM, PL, CCB	Software System Test Plan
		Software System Test Plan
		Review Documentation
(11) Release	Project Lead	Baseline release to Configuration Control
		Inspection Report
(12) Maintenance	Refer below (Step 12)	Refer below (Step 12)

Step (1)

Software requirements define the problem to be solved by analyzing the customer requirements of the solution in a software requirements specification (SRS). The requirements portion of the life cycle is "the period of time in the software life cycle during which the requirements for a software product are defined and documented" [6].

Customer requirements should be in the form of a statement of work, tasking statement, or software system specification (SSS).

The associated Project Lead, Program Manager, and any associated CCB will review the SRS for technical content and clarity. The SRS will be signed by the [project abbreviation] Program Manager and a [customer name] representative after joint review. This constitutes their agreement on what is to be produced. The Project Lead, or designee, will review the SRS for adherence to the documentation template provided in Appendix A of the [project abbreviation] Software Requirements Management Plan. The SQA Lead will ensure that this review has been completed.

Software project/development planning follows software requirements in the software development life cycle. Software project/development planning requires the documentation of a proposed project's schedule, resources, and manpower estimates. This software project plan will be reviewed for technical content and clarity by the associated Project Lead (PL), Program Manager, senior [company name] management, and if appropriate, the associated CCB. The customer will also participate in plan review and authorize its implementation.

The PL, or a designee, will also review the software project plan for adherence to the documentation template provided for the [company name]. The [project abbreviation] Program Manger and a representative from [customer name] will review actual progress against existing plans to ensure tracking efficiency. These reviews will occur no less than bi-monthly. The PM will report progress tracking to [company name] senior management monthly.

Note: It is important to carefully consider document review coordination time and repair time when project planning.

Step (2)

Software system test planning follows software requirements development in defining how the final product will be tested and what constitutes acceptable results. This step provides the basis for both acceptance and system testing. Issues identified during software test planning [also referred to as developmental test and evaluation (DT&E) test planning] may result in changes to the software requirements. A highly desirable by-product of system test planning is the further refinement of the software requirements. Errors found early in the software development process are less costly and easier to correct.

The software test plan will be reviewed for technical content and clarity by the associated Project Lead, Program Manager, and, if appropriate, the associated CCB. The customer will determine, and [company name] will agree to, test tolerances.

The Project Lead, or a designee, will review the test plan to ensure that all requirements items identified in the SRS are also identified in the software test plan.

Step (3)

Preliminary design breaks the specified problem into manageable components, creating the architecture of the system to be built. The architecture in this step is refined until the lowest level of software components or CSUs are created. The CSUs will represent the algorithms to be used in fulfilling the customer's needs. The preliminary design will be documented as a draft software design document.

The PL will review any existing test plan to ensure that all requirements identified in the SRS are also identified in the software design document. The SQA Lead will ensure that this review has been completed.

Step (4)

Integration test planning follows preliminary design by defining how the components of any related system architecture will be brought back together and successfully integrated. Issues identified during this step will help to define the ordering of component development. Other issues identified during this step include recognition of interfaces that may be difficult to implement. This saves on the final cost by allowing for early detection of a problem. Integration testing will be included as part of the software (DT&E) test plan.

This plan will be reviewed for technical content and clarity by the associated Project Lead, Program Manager, and, if appropriate, the CCB. The customer will determine and [company name] will agree to test tolerances. The Project Lead, or a designee will review the test plan to ensure that all requirements identified in the SDD are also identified in the software system test plan. The SQA Lead will ensure that this review has been completed.

Step (5)

Detailed design defines the algorithms used within each of the CSUs. The Design is "the period of time in the software life cycle during which the designs for architecture, software components, interfaces, and data are created, documented, and verified to satisfy requirements" [2].

The detailed design will be documented as the final version of the software design document (SDD). The software design document will be reviewed for technical content and clarity by the associated Project Lead, Program Manager, and, if appropriate, the [project abbreviation] CCB. The customer will also participate in SDD review and authorize its implementation.

Step (6)

Unit test planning follows the detailed design by creating the test plan for the individual CSUs. Of primary importance in this step is logical correctness of algorithms and correct handling of problems such as bad data. As in all other test planning steps, feedback to detailed design is essential. Unit testing will be included as part of the software (DT&E) test plan.

The associated Project Lead, Program Manager, and the CCB will review this plan for technical content and clarity. The customer will determine and [company name] will agree to test tolerances. The PL will review the test plan to ensure that all requirements items identified in the SRS are also identified in the Software (DT&E) Test Plan. The SQA Lead will ensure that this review has been completed.

Step (7)

Coding is the activity of implementing the system in a machine executable form. Implementation is "the period of time in the software life cycle during which a software product is created from documentation and debugged" [2].

The PL, or a designee, using existing coding standards to ensure compliance with design and requirements specifications, will review the implemented code. Periodic program SQA reviews will be conducted to ensure that code reviews are being properly implemented. The results of these reviews will comprise a "lessons learned" document. Results will also be reported to senior [company name] management.

Step (8)

Unit testing executes the tests created for each CSU. This level of test will focus on the algorithms with equal emphasis on both "white" and "black" box test cases. White box testing is testing that is based on an understanding of the internal workings of algorithms. Black box testing focuses on final results and integrated software operation.

The PL or a designee will "walk through" software test procedures, ensuring that results are recorded and procedures are documented and based on the evaluation of documented requirements.

Step (9)

Integration testing executes the integration strategy by orderly combining CSUs into higher-level components until a full system is available. This level of test will focus on the interface.

The associated Software/DT&E test plan will document system testing procedures and standards.

Step (10)

System testing executes the entire system to ensure that the requirements have been satisfied completely and accurately. This level of testing will focus on functionality and is primarily black-box testing. Black-box testing is focused on the functionality of the product (determined by the requirements); therefore, testing is based on input/output knowledge rather than processing and internal decisions.

The associated Software/DT&E test plan will document integration testing procedures and standards.

Step (11)

Release is the entry of the new or upgraded software into configuration management and the replacement of the previous executable with the upgraded executable, as appropriate. Please refer to Appendix C for the checklists used in SQA reviews of releasable items.

Step (12)

Maintenance is the receipt of and response to problems encountered by the user, and upgrades requested by the customer. Each problem or upgrade will spawn a new SQA process. (Refer to Figure C-5.)

The maintenance life cycle of [PROJECT ABBREVIATION] software consists of the following stages shown in Figure C-5 and tabularly in Table C-23.

The software change process begins with the creation of a change request and is controlled by the software configuration management process for each program. For this discussion, change enhancement requests (CERs) will include software problem reports (SPRs), baseline change requests (BCRs), internal test reports

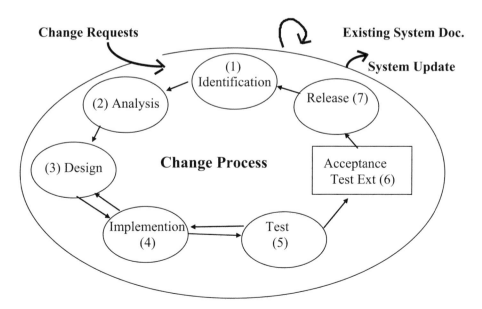

Figure C-5. Software maintenance life cycle.

Table C-23. Software maintenance life cycle products

Life cycle activity	Review by	Product
(1) Identification	PL, PM	CER Categorization
		SCM Audit
(2) Analysis	PL, PM, CCB, [Customer Name]	CER Scheduling
		Functional Test Review
(3) Design	PL, PM, CCB, [Customer Name]	Design Review
(4) Implement	PL, PM, SQAM	Walk-through Report
		Audit Documentation
(5) Test	PL, PM, SQAM	Test Report
		Audit Documentation
(7) Release	PL, PM	Staff Meeting Minutes

(ITRs) and interface change notices (ICNs). These forms are shown in the [PROJECT ABBREVIATION] Software Configuration Management Plan.

The steps in the software maintenance process are described below.

Step (1). Identification and categorization of the CER. This step ensures the change request is unique (not previously submitted), is valid (not the result of operator error or misunderstanding), and is within the scope of the system. Categorization is concerned with the criticality of the change (urgent versus routine) and the type of change (corrective or enhancement).

The Project Lead assumes the initial responsibility for the determination of whether a CER is valid or invalid. Valid change requests require further analysis [Step (2)]. The SQAM will conduct periodic reviews of the software configuration management tracking procedures used for the identification, tracking, and prioritization of CERs.

Step (2). Analysis examines the source and reason for the change. The CERs will be reviewed for technical validity and project impact by the associated Project Lead, Program Manager, and, if appropriate, the associated CCB. The customer may also participate in CER review and authorize their implementation. Corrective changes need to be analyzed to find the true source of the problem; enhancements need to be analyzed to determine the scope of the change.

Note: enhancements are normally additions to the system that entail new or modified requirements. This is similar to starting at the top of Figure C-5. Part of this step is determining the change to functional testing that will be used upon completion of the change.

Those CERs identified as valid are prioritized and scheduled by the Project Lead, the Program Manager, the Customer, and associated CCBs.

Step (3). Design retrofits changes into the existing architecture and algorithms. As in development, the integration of changed CSUs and execution of algorithms must be planned early, to catch problems before testing occurs.

The SQAM will conduct periodic reviews to ensure that the Project Lead reviews designs for technical appropriateness. Results of these reviews will be reported to the Program Manager.

Step (4). Implementation converts new solution into machine executable form. The implemented code will be reviewed using existing coding standards and for compliance to design and requirements specifications. The code will be reviewed by the implementer and then by the PL, or a designee.

Periodic SQA reviews will be conducted to ensure that code reviews are being implemented. Results of these reviews will be reported to the Program Manager.

Step (5). Testing in this process is the combination of unit, integration, and system testing, as described during development. This testing process is often called regression testing due to the need to modify and repeat testing previously performed. A critical issue at this point is the need for repeatability of testing. Changes are vital to the life of the system, as is proper testing; therefore, the testing process should be developed for reuse.

The Project Leads are responsible for ensuring that all CERs scheduled for a production cycle are implemented in the software released for acceptance testing. The SQAM will conduct periodic reviews of the SCM procedures being implemented. Changes to the software baseline will be traced back to the initiating CERs reporting any anomalies.

Step (6). Acceptance testing is done to achieve the customer's agreement on completion of the change. Acceptance testing is controlled by the customer for [PROJECT ABBREVIATION] software products, both maintenance and development.

Step (7). Release processes should be considered during identification of the change and follow an established pattern for the customer. Periodic releases of "builds" are easier to manage and provide more consistency; independent releases provide for quicker responses to changing demands. There are considerable trade-offs between the two extremes that should be negotiated with the customer to reach the best mix of speed and control.

The customer and/or associated representative determines both the "when" and "recipient" of a software release. The PM will, at a minimum, conduct bi-weekly staff reviews that will include review of release status and CER implementation progress. The SQA Lead will ensure that these reviews occur as described in the associated software project schedule.

Minimum Set of Software Reviews [3]

Table C-24 presents a list of the recommended minimum set of software reviews as described in IEEE Std 730, IEEE Standard for Software Quality Assurance Plans.

Table C-24. Minimum set of software reviews

Software Specification Review
Architectural Design Review
Detailed Design Review
Verification and Validation Plan Review
Functional Audit
Physical Audit
In-Process Audit
Managerial Reviews
Software Configuration Management Plan Review
Post Implementation Review

The list of reviews provided here is not an absolute. Rather, these are provided as an example of what is contained in this standard. Each organization must determine its own recommended minimum set of software reviews.

SQA Inspection Log

An inspector uses the SQA inspection log (Figure C-6) form during his/her review of the inspection package materials. The recorder may also use it during an inspection. In preparing for an inspection, an inspector uses this form to identify all defects found. During an inspection, the recorder uses this form to record all official defects identified. An electronic version of this log form is provided on the companion CD-ROM, entitled *SQA Inspection Log.doc.*

Inspection Log Description

Log number. The log or tracking number assigned to this inspection.

Date. During an inspection, the recorder enters the date on which the inspection occurred. The date block can be ignored by an inspector using this form during the preinspection review.

Software item. The name of the software item being inspected.

Inspector. The inspector block can be left blank when used by the recorder during an inspection. When using this form during a preinspection review, the inspector name is entered in this block.

Moderator. The moderator's name.

Author. The name of the author responsible for the inspection package is entered in this block.

Review type. The type of inspection being conducted is entered in this block.

Category. A category is a two-character code that specifies the class and severity of a defect. A category has the format XY, where X is the severity (**M**ajor or m**I**nor, see below) and Y is the class.

Severity:

Major = a defect that would cause a problem in program operation

m**I**nor = all other defects

Class:

M(issing) = required item is missing from software element

S(tandards Compliance) = nonconformance to existing standards

W(rong) = an error in a software element

E(xtra) = unneeded item is included in a software element

A(mbiguous) = an item of a software element is ambiguous

I(nconsistent) = an item of a software element is inconsistent

Log number: _____

Date: _____

Software item: _____

Inspector: _____

Moderator: _____

Author: _____

Time spent on review: _____

Review type: _____

Location	Defect Description	Category

Figure C-6. SQA inspection log form.

VERIFICATION

Inspection Log Defect Summary

The inspection log defect summary shown in Figure C-7 is provided as an example. This may used to record a summary of all defects identified during the inspection process. Once the inspection team has agreed to a disposition on the inspection, the leader should complete this form. An electronic copy of this log, *Inspection Log Defect Summary.doc,* is provided on the companion CD-ROM.

Inspection Log Defect Summary Description

Log number. The recorder will enter the log number assigned to this inspection.

Date. The recorder will enter the date of the inspection in this block.

Software item. The recorder will enter the name of the software item that is being inspected.

Moderator. The name of the inspection moderator.

Inspection type. The recorder will enter what type of inspection that is being conducted in this block.

Defects found. The recorder will record counts of all defects found during review of the inspection package. The defect counts will be recorded in the table as follows:

- The Insp. column will contain name of the inspector.
- The Major Defects columns (M, S, W, E, A, and I) will hold identifiers that correspond to the number of major defects found for a specific type in one of the six possible classes. As an example, if the "W" column held the number 3, then it would mean that there were 3 major defects found that were classed as Wrong.
- The first Total column will hold a total of all the major defects found for a type.
- The Minor Defects columns (M, S, W, E, A, and I) will hold numbers that correspond to the number of minor defects found for a specific type in one of the five possible classes. As an example, if the "A" column held the number 3, then it would mean that there were 3 minor defects found that were classed as Ambiguous.
- The second Total column will hold a total of all the minor defects found for a type.

Inspection Report

The inspection report form shown in Figure C-8 is provided as an example. This may used by the inspection leader to officially close the inspection. Once the in-

Log number: _____   Date: _____

Software item: _____

Moderator: _____   Review type: _____

Defects found:

Insp.	Major Defects							Minor Defects						
	M	S	W	E	A	I	TOTAL	M	S	W	E	A	I	TOTAL

Legend:
   Major = a defect that would cause a problem in program operation
   Minor = all other defects
   M(issing) = required item is missing from software element
   S(tandards Compliance) = nonconformance to existing standards
   W(rong) = an error in a software element
   E(xtra) = unneeded item is included in a software element
   A(mbiguous) = an item of a software element is ambiguous
   I(nconsistent) = an item of a software element is inconsistent

Figure C-7. Inspection log defect summary.

Log number: _____   Date: _____

Software item: _____

Inspection type: _____   Duration: _____

Number of inspectors: _____   Total prep time: _____

Size of materials inspected: _____
_____
_____

Disposition: Accept ____   Conditional ____   Reinspect ____

Estimated rework effort: _____ (Hours)
Rework to be completed by: _____
Reinspection scheduled for: _____

Inspectors:
_____   _____
_____   _____

Author(s):
_____   _____

Recorder: _____

Leader: _____

Leader certification: _____ Date: _____

Additional comments: _____

Figure C-8. Inspection report form.

spection team has agreed to a disposition on an inspection, the inspection leader should complete this form. An electronic version of this report, *Inspection Report.doc,* is provided on the companion CD-ROM.

Inspection Report Description

Log number. The log or tracking number assigned to this inspection.
Date. The date of the inspection.
Software item. The name of the software item being inspected.
Inspection type. The type of inspection being conducted.
Duration. The duration of the inspection.
Number of inspectors. The number of inspectors present during the inspection.

Total Prep time. The total preparation time spent in preparation for the inspection.

Size of materials. The size of the materials inspected.

Disposition. The disposition of the inspection.

Estimated rework effort. The estimated rework effort time, in hours.

Rework to be completed by. The estimated date of rework completion. If no rework is needed, enter "N/A."

Reinspection scheduled for. The date for a reinspection of the software item to ensure rework completeness. It is the moderator's responsibility to decide if a reinspection is needed. If it is not needed, this block contains "N/A."

Inspectors. The names of the inspectors present at the inspection.

Author(s). The name(s) of the author(s) responsible for the software item.

Recorder. The name of the recorder present at the inspection.

Leader. The leader's name in this block.

Leader certification. The leader will sign and date this block to signal that the inspection is concluded.

Additional comments. The leader will enter additional comments as necessary for clarification.

Requirements Walk-through Form

The form in Figure C-9 is provided as an example walk-through checklist used in support of the peer review of software or system requirements. This type of form may be used to document issues and resolutions when performing project walk-throughs.

Software Project Plan Walk-through Checklist

The form in Figure C-10 is provided as an example walk-through checklist used in support of the peer review of a software project plan. This type of form may be used to document issues and resolutions when performing project walk-throughs.

Preliminary Design Walk-through Checklist

The form shown in Figure C-11 is provided as an example walk-through checklist used in support of the peer review of preliminary design. This type of form may be used to document issues and resolutions when performing project walk-throughs.

Software project:

Software item(s): _____

Review date: _____

Reviewer(s):

_____   _____

_____   _____

_____   _____

_____   _____

Moderator, please respond to the following:

	Yes	No*
1) Is there a clear understanding of the problem that this system is designed to solve?	____	____
2) Are the external and internal interfaces properly defined?	____	____
3) Are all the requirements traceable to the system level?	____	____
4) Is prototyping conducted for customer?	____	____
5) Is the performance achievable with constraints imposed by other system elements?	____	____
6) Are schedule, resources, and budget consistent with requirements?	____	____
7) Are validation criteria complete?	____	____

*If "No" was the answer to any of the above, please attach problems found and suggestions to improve.

Please attach additional comments and a copy of requirement items.

Figure C-9. Requirements walk-through checklist.

Detailed Design Walk-through Checklist

The form shown in Figure C-12 is provided as an example walk-through checklist used in support of the peer review of detailed design. This type of form may be used to document issues and resolutions when performing project walk-throughs.

Program Code Walk-through Checklist

The form in Figure C-13 is provided as an example walk-through checklist used in support of the peer review of developed code. This type of form may be used to document issues and resolutions when performing project walk-throughs. Note: the checklist assumes that a design walk-through has already been done.

Software project:

Software item(s): _____

Review date: _____

Reviewer(s):

_____   _____
_____   _____
_____   _____
_____   _____

Moderator, please respond to the following:

	Yes	No*
1) Is the software scope unambiguously defined and bounded?	____	____
2) Is the terminology clear?	____	____
3) Are the resources adequate for the scope?	____	____
4) Are the resources readily available?	____	____
5) Are the tasks properly defined and sequenced?	____	____
6) Is task parallelism reasonable given the available resources?	____	____
7) Have both historical productivity and quality data been used?	____	____
8) Have any differences in estimates been reconciled?	____	____
9) Are the preestablished budgets and deadlines realistic?	____	____
10) Is the schedule consistent?	____	____

*If "No" was the answer to any of the above, please attach problems found and suggestions to improve.

Please attach additional comments and a copy of requirement items.

Figure C-10. Software project plan walk-through checklist.

Test Plan Walk-through Checklist

form shown in Figure C-14 is provided as an example walk-through checklist used in support of the peer review of a software test plan. This type of form may be used to document issues and resolutions when performing project walk-throughs.

Walk-through Summary Report

The form shown in Figure C-15 is provided as an example walk-through summary report. This may be used to document issues and resolutions when performing peer reviews.

Software project:

Software item(s): _____

Review date: _____

Reviewer(s):

_____   _____
_____   _____
_____   _____
_____   _____

Moderator, please respond to the following:

	Yes	No*
1) Are software requirements reflected in the software architecture?	____	____
2) Is effective modularity achieved? Are modules functionally independent?	____	____
3) Is the program architecture factored?	____	____
4) Are interfaces defined for modules and external system elements?	____	____
5) Is the data structure consistent with information domain?	____	____
6) Is the data structure consistent with software requirements?	____	____
7) Has maintainability been considered?	____	____

*If "No" was the answer to any of the above, please attach problems found and suggestions to improve.

Please attach additional comments and a copy of requirement items.

Figure C-11. Preliminary design walk-through checklist.

Classic Anomaly Class Categories

Anomaly classes provide evidence of nonconformance and may be categorized. The following examples are provided:

Missing

Extra (superfluous)

Ambiguous

Inconsistent

Improvement desirable

Not conforming to standards

Risk-prone, that is, the review finds that, although an item was not shown to be "wrong," the approach taken involves risks (and there are known safer alternative methods)

Software project:

Software item(s): _____

Review date: _____

Reviewer(s):

_____   _____

_____   _____

_____   _____

_____   _____

Moderator, please respond to the following:

	Yes	No*
1) Does the algorithm accomplish the desired function?	___	___
2) Is the algorithm logically correct?	___	___
3) Is the interface consistent with the architectural design?	___	___
4) Is logical complexity reasonable?	___	___
5) Has error handling been specified and built in?	___	___
6) Is local data structure properly defined?	___	___
7) Are structured programming constructs used throughout?	___	___
8) Is design detail amenable to the implementation language?	___	___
9) Which are used: operating system or language-dependent features?	___	___
10) Has maintainability been considered?	___	___

*If "No" was the answer to any of the above, please attach problems found and suggestions to improve.

Please attach additional comments and a copy of requirement items.

Figure C-12. Detailed design walk-through checklist.

Factually incorrect

Not implementable (e.g., because of system constraints or time constraints)

Editorial

Validation

Example Test Classes

Check for Correct Handling of Erroneous Inputs

Test Objective. Check for proper handling of erroneous inputs: characters that are not valid for this field, too many characters, not enough characters, value too large, value too small, all selections for a selection list, no selections, all mouse buttons clicked or double clicked all over the client area of the item with focus. Test class # xx.

Software project:

Software item(s): _____

Review date: _____

Reviewer(s):

_____   _____

_____   _____

_____   _____

_____   _____

Moderator, please respond to the following:

	Yes	No*
1) Is design properly translated into code?		
2) Are there misspellings or typos?	___	___
3) Has proper use of language conventions been made?	___	___
4) Is there compliance with coding standards for language style, comments, and module prologue?	___	___
5) Are incorrect or ambiguous comments present?		
6) Are typing and data declaration proper?	___	___
7) Are physical contents correct?	___	___
8) Have all items on the design walk-through checklist been reapplied (as required)?	___	___

*If "No" was the answer to any of the above, please attach problems found and suggestions to improve.

Please attach additional comments and a copy of requirement items.

Figure C-13. Program code walk-through checklist.

Validation Methods Used. Test.

Recorded Data. User action or data entered, screen/view/dialog/control with focus, resulting action.

Data Analysis. Was resulting action within general fault-handling-defined capabilities in the [PROJECT ABBREVIATION] SysRS and design in [PROJECT ABBREVIATION] SDD?

Assumptions and Constraints. None.

Check for Maximum Capacity

Test Objective. Check software and database maximum capacities for data. Enter data until maximum number of records specified in the design is reached for each table, operate program, and add one more record. Test class # xx.

Validation Methods Used. Test.

Software project:

Software item(s): _____

Review date: _____

Reviewer(s):

_____   _____
_____   _____
_____   _____
_____   _____

Moderator, please respond to the following:

	Yes	No*
1) Have major test activities been properly identified and sequenced?	___	___
2) Has traceability to validation criteria/requirements been established as part of software requirements analysis?	___	___
3) Are major functions demonstrated early?		
4) Is the test plan consistent with the overall project plan?	___	___
5) Has a test schedule been explicitly defined?	___	___
6) Are test resources and tools identified and available?	___	___
7) Has a test record-keeping mechanism been established?	___	___
8) Have test drivers and stubs been identified, and has work to develop them been scheduled?	___	___
9) Has stress testing for software been specified?	___	___

*If "No" was the answer to any of the above, please attach problems found and suggestions to improve.

Please attach additional comments and a copy of requirement items.

Figure C-14. Test plan walk-through checklist.

Recorded Data. Record number of records in each table, resulting actions.

Data Analysis. Was resulting action to the maximum plus one normal?

Assumptions and Constraints. This test requires someone to create through some method a populated database with records several times more than what actually exists in the sample data set. This takes a good deal of time. Integration and qualification test only.

User Interaction Behavior Consistency

Test Objective. Is the interaction behavior of the user interface consistent across the application or module under test: tab through controls, using mouse click and double click on all controls and in null area, maximize, minimize, normalize, switch focus to another application and then back, update data on one view that is included on another and check to see if the other view is updated

Project name:	Date:
Element reviewed:	
Review team:	
Problems found:	
Proposed solutions:	
Comments:	
Action taken:	

Figure C-15. Walk-through summary report.

when data is saved on the first view, use function keys and movement keys and other standard key combinations (clipboard combos, control key windows, and program defined sets, Alt key defined sets), enter invalid control and Alt key sets to check for proper handling. Test class # xx.

Validation Methods Used. Test, inspection.

Recorded Data. Record any anomalies of action resulting from user action not conforming to the behavioral standards for Windows programs.

Data Analysis. Was resulting action within behavioral standards of windows programs as defined in the [PROJECT ABBREVIATION] SysRS and design in [PROJECT ABBREVIATION] SDD? Was behavior consistent across the application or module as defined in the [PROJECT ABBREVIATION] SysRS and design in [PROJECT ABBREVIATION] SDD?

Assumptions and Constraints. If testing at module level, the multiple view portion of the test may not apply due to having only a single view.

Retrieving Data

Test Objective. Is the data retrieved correct? For each dialog, list box, combo box and other controls that show lists, check the data displayed for correctness. Test class # xx.

Validation Methods Used. Test, inspection.

Recorded Data. Record data displayed and data sources (records from tables, resource strings, code sections).

Data Analysis. Was data displayed correctly? Compare data displayed with sources.

Assumptions and Constraints. Requires alternate commercial database software to get records from the database.

Saving Data

Test Objective. Is the data entered saved to the database correctly? For each dialog, list box, combo box and other controls that show lists, check the data entered and saved for correctness in the database. Test class # xx.

Validation Methods Used. Test, inspection.

Recorded Data. Record data entered and data destinations (records from tables).

Data Analysis. Was data saved correctly? Compare data entered with destination.

Assumptions and Constraints. Requires alternate commercial database software to get records from the database.

Display Screen and Printing Format Consistency

Test Objective. Is user interface screens organized and labeled consistently, are printouts formatted as specified? Enter data to maximum length of field in a printout and then print, show all screens (views, dialogs, print previews, OLE devices) and dump their image to paper. Test class # xx.

Validation Methods Used. Inspection.

Recorded Data. Screen dumps and printouts.

Data Analysis. Was the printout format correct? Were the fields with max length data not clipped? Were the labels and organization of screens consistent across the application or module as defined in the [PROJECT ABBREVIATION] SDD?

Assumptions and Constraints. The module that performs forms printing is required with all other modules during their testing.

Check Interactions between Modules

Test Objective. Check the interactions between modules. Enter data and save it in one module and switch to another module that uses that data to check for latest data entered. Switch back and forth between all of the modules that will be manipulating data and check for adverse results or program faults. Test class # xx.

Validation Methods Used. Demonstration.

Recorded Data. Screen dumps.

Data Analysis. Were resulting actions within specifications as defined in the [PROJECT ABBREVIATION] SysRS and design in [PROJECT ABBREVIATION] SDD?

Assumptions and Constraints. Requires customer participation. Requires all modules and supporting software.

Measure Time of Reaction to User Input

Test Objective. Check average response time to user input action: clock time from (saves, retrieves, dialogs open and closes, views open and closes), clock time from any response to user action that takes longer than 2 seconds. Test class # xx.

Validation Methods Used. Test, analysis.

Recorded Data. Record action and response clock time. Organize into categories and average their values. Are all average values less than the minimum response time specified?

Data Analysis. Organize into categories and average their values. Are all average values less than the minimum response time specified as defined in the [PROJECT ABBREVIATION] SysRS and design in [PROJECT ABBREVIATION] SDD?

Assumptions and Constraints. None.

Functional Flow

Test Objective. Exercise all menus, buttons, hotspots, and so on that cause a new display (view, dialog, OLE link) to occur. Test class # xx.

Validation Methods Used. Demonstration.

Recorded Data. Screen dumps.

Data Analysis. Were resulting actions with specifications as defined in the [PROJECT ABBREVIATION] SysRS and design in [PROJECT ABBREVIATION] SDD?

Assumptions and Constraints. Requires customer participation. Requires all modules and supporting software.

Examples of System Testing

Functional Testing . Testing against operational requirements is functional testing. This section should provide references to the SysRS to show a complete list of functional requirements and resulting test strategy.

Performance Testing. Testing against performance requirements is performance testing. This section should provide references to the SysRS to show a complete list of performance requirements and resulting test strategy.

Reliability Testing. Testing against reliability requirements is reliability testing. This section should provide references to the SysRS to show a complete list of reliability requirements and resulting test strategy.

Configuration Testing. Testing under different hardware and software configurations to determine an optimal system configuration is configuration testing. This section should provide references to the SysRS to show a complete list of configuration requirements and resulting test strategy.

Availability Testing. Testing against its operational availability requirements is availability testing. This section should provide references to the SysRS to show a complete list of availability requirements and resulting test strategy.

Portability Testing. Testing against portability requirements is portability testing. This section should provide references to the SysRS to show a complete list of portability requirements and resulting test strategy.

Security and Safety Testing. Testing against security and safety requirements is security and safety testing. This section should provide references to the SysRS to show a complete list of security and safety requirements and resulting test strategy.

System Usability Testing. Testing against usability requirements is system usability testing. This section should provide references to the SysRS to show a complete list of system usability requirements and resulting test strategy.

Internationalization Testing. Testing against internationalization requirements is internationalization testing. This section should provide references to the SysRS to show a complete list of internationalization requirements and resulting test strategy.

Operations Manual Testing. Testing against the procedures in the operations manual to see if it can be operated by the system operator is operations manual testing. This section should provide references to the operations manual to show a complete list of operator requirements and resulting test strategy.

Load Testing. Testing that attempts to cause failures involving how the performance of a system varies under normal conditions of utilization is load testing. This section should provide the test strategy.

Stress Testing. Testing that attempts to cause failures involving how the system behaves under extreme but valid conditions (e.g., extreme utilization, insufficient memory inadequate hardware, and dependency on overutilized shared resources) is stress testing. This section should provide the test strategy.

Robustness Testing. Testing that attempts to cause failures involving how the system behaves under invalid conditions (e.g., unavailability of dependent applications, hardware failure, and invalid input such as entry of more than the maximum amount of data in a field) is robustness testing. This section should provide the test strategy.

Contention Testing. Testing that attempts to cause failures involving concurrency is contention testing. This section should provide the test strategy.

Test Design Specification

This example test design specification is based upon IEEE-Std 829, IEEE Standard for Software Test Documentation [5].

Purpose

This section should provide an overview description the test approach and identify the features to be tested by the design.

Outline

A test design specification should have the following structure:

1. Test design specification identifier. Specify the unique identifier assigned to this test design specification as well as a reference to any associated test plan.
2. Features to be tested. Identify all test items, describing the features that are the targets of this design specification. Supply a reference for each feature to any specified requirements or design.
3. Approach refinements. Specify any refinements to the approach and any specific test techniques to be used. The method of analyzing test results should be identified (e.g., visual inspection). Describe the results of any analysis that provides a rationale for test case selection. For example, a test case used to determine how well erroneous user inputs are handled.
4. Test identification. Include the unique identifier and a brief description of each test case associated with this design. A particular test case may be associated with more than one test design specification.
5. Feature pass/fail criteria. Specify the criteria to be used to determine whether the feature has passed or failed.

The sections should be ordered in the specified sequence. Additional sections may be included at the end. If some or all of the content of a section is in another document, then a reference to that material may be listed in place of the corresponding content. The referenced material must be attached to the test design specification or readily available to users.

Test Case Specification

This example test case specification is based upon IEEE-Std 829, IEEE Standard for Software Test Documentation [5].

Purpose

To define a test case identified by a test design specification.

Outline

A test case specification should have the following structure:

1. Test case specification unique identifier. Specify the unique identifier assigned to this test case specification.
2. Test items. Identify and briefly describe the items and features to be exercised by this test case. For each item, consider supplying references to the following test item documentation: requirements specification, design specification, users guide, operations guide, and installation guide.
3. Input specifications. Specify each input required to execute the test case. Some of the inputs will be specified by value (with tolerances where appropriate), whereas others, such as constant tables or transaction files, will be specified by name. Specify all required relationships between inputs (e.g., timing).
4. Output specifications. Specify all of the outputs and features (e.g., response time) required of the test items. Provide the exact value (with tolerances where appropriate) for each required output or feature.
5. Environmental needs. Specify all hardware and software requirements associated with this test case. Include facility or personnel requirements if appropriate.
6. Special procedural requirements. Describe any special constraints on the test procedures that execute this test case.
7. Intercase dependencies. List the identifiers of test cases that must be executed prior to this test case. Summarize the nature of any dependencies.

If some or all of the content of a section is in another document, then a reference to that material may be listed in place of the corresponding content. The referenced material must be attached to the test case specification or readily available to users of the case specification.

A test case may be referenced by several test design specifications and enough specific information must be included in the test case to permit reuse.

Test Procedure Specification

This example test procedure specification is based upon IEEE-Std 829, IEEE Standard for Software Test Documentation [5].

Purpose

To document the steps in support of test set execution or, more generally, the steps used to evaluate a set of features associated with a software item.

Outline

A test procedure specification should have the following structure:

1. Test procedure specification identifier. The unique identifier assigned to the test procedure specification. A reference to the associated test design specification should also be provided.

2. Purpose. Describe the purpose of this procedure, providing a reference for each executed test case. In addition, provide references to relevant sections of the test item documentation (e.g., references to usage procedures) where appropriate.

3. Special requirements. Identify any special requirements that are necessary for the execution. These may include prerequisite procedures, required skills, and environmental requirements.

4. Procedure steps. Describe the procedural steps supporting the test in detail. The following provides guidance regarding the level of detail required:

Log. Describe the method or format for logging the results of test execution, the incidents observed, and any other test events.

Setup. Describe the sequence of actions necessary to prepare for execution of the procedure.

Start. Describe the actions necessary to begin execution of the procedure.

Proceed. Describe any actions necessary during execution of the procedure.

Measure. Describe how the test measurements will be made.

Shutdown. Describe the actions necessary to immediately suspend testing if required.

Restart. Identify any restart points and describe the actions necessary to restart the procedure at each of these points.

Stop. Describe the actions necessary to bring execution to an orderly halt.

Wrap up. Describe the actions necessary to restore the environment.

Contingencies. Describe the actions necessary to deal with anomalous events that may occur during execution.

The sections should be ordered as described. Additional sections, if required, may be included at the end of the specification. If some or all of the content of a section is in another document, then a reference to that material may be listed in place of the corresponding content. The referenced material must be attached to the test procedure or readily available to users of the procedure specification.

Test Item Transmittal Report

This example test item transmittal report is based upon IEEE-Std 829, IEEE Standard for Software Test Documentation [5].

Purpose

To identify all items being transmitted for testing. This report should identify the person responsible for each item, the physical location of the item, and item status. Any variations from the current item requirements and designs should be noted in this report.

Outline

A test item transmittal report should have the following structure:

1. Transmittal reports identifier. The unique identifier assigned to this test item transmittal report.
2. Transmitted items. Describe all test items being transmitted, including their version/revision level. Supply references associated with the documentation of the item including, but not necessarily limited to, the test plan relating to the transmitted items. Indicate the individuals responsible for the transmitted items.
3. Location. Identify the location of all transmitted items. Identify the media that contain the items being transmitted. Indicate media identification and labeling.
4. Status. Describe the status of the test items being transmitted, listing all incident reports relative to these transmitted items.
5. Approvals. List all transmittal approval authorities, providing a space for associated signature and date.

The sections should be ordered as described. Additional sections, if required, may be included at the end of the specification. If some or all of the content of a section is in another document, then a reference to that material may be listed in place of the corresponding content. The referenced material must be attached to the test item transmittal report or readily available to users of the transmittal report.

Test Log

This example test log is based upon IEEE-Std 829, IEEE Standard for Software Test Documentation [5].

Purpose

To document the chronological record of test execution.

Outline

A test log should have the following structure:

1. Test log identifier. The unique identifier assigned to this test log.
2. Description. This section should include the identification of the items being tested, environmental conditions, and type of hardware being used.
3. Activity and event entries. Describe event start and end activities, recording the occurrence date, time, and author. A description of the event execution, all results, associated environmental conditions, anomalous events, and test incident report number should all be included as supporting information.

The sections should be ordered in the specified sequence. Additional information may be included at the end. If some or all of the content of a section is in another document, then a reference to that material may be listed in place of the corresponding content. The referenced material must be attached to the test log or readily available to users of the log.

Test Incident Report

This example test incident report is based upon IEEE-Std 829, IEEE Standard for Software Test Documentation [5].

Purpose

To document all events occurring during the testing process requiring further investigation.

Outline

A test incident report should have the following structure:

1. Test incident report identifier. The unique identifier assigned to this test incident report.
2. Summary. Provide a summary description of the incident, identifying all test items involved and their version/revision level(s). References to the appropriate test procedure specification, test case specification, and test log should also be supplied.
3. Incident description. Provide a description of the incident. This description should include the following items: inputs and anticipated results, actual results, all anomalies, the date and time and procedure step, a description of the environment, any attempt to repeat, and the names of testers and observers.
4. Related activities and observations that may help to isolate and correct the cause of the incident should be included (e.g., describe any test case executions that might have a bearing on this particular incident and any variations from the published test procedure).
5. Impact. Describe the impact this incident will have on existing test plans, test design specifications, test procedure specifications, or test case specifications.

The sections should be ordered in the sequence described above. Additional sections may be included at the end if desired. If some or all of the content of a section is in another document, then a reference to that material may be listed in place of the corresponding content. The referenced material must be attached to the test incident report or made readily available to users of the incident report.

Test Summary Report

This example test summary report is based upon IEEE-Std 829, IEEE Standard for Software Test Documentation [5].

Purpose

To provide a documented summary of the results of testing activities and associated result evaluations.

Outline

A test summary report should have the following structure:

Test summary report identifier. Include the unique identifier assigned to this test summary report.

Summary. Provide a summary of items tested. This should include the identification of all items tested, their version/revision level, and the environment in which the testing activities took place. For each test item, also supply references to the following documents if they exist: test plan, test design specifications, test procedure specifications, test item transmittal reports, test logs, and test incident reports.

Variances. Report any variances of the test items from their specifications associating the reason for each variance.

Comprehensiveness assessment. Identify any features not sufficiently tested, as described in the test plan, and explain the reasons.

Summary of results. Provide a summary of test results, including all resolved and unresolved incidents.

Evaluation. This overall evaluation should be based upon the test results and the item-level pass/fail criteria. An estimate of failure risk may also be included.

Summary of activities. Provide a summary of all major testing activities and events.

Approvals. List the names and titles of all individuals who have approval authority. Provide space for the signatures and dates.

The sections should be ordered in the specified sequence. Additional sections may be included just prior to approvals if required. If some or all of the content of a section is in another document, then a reference to that material may be listed in place of the corresponding content. The referenced material must be attached to the test summary report or readily available to users.

JOINT REVIEW

Open Issues List

The form in Figure C-16 is presented as an example that may be used in support of the documentation and tracking of issues associated with a software project. These issues are typically captured during reviews or team meetings and should be reviewed on a regular basis. An electronic version of this work product is provided on the companion CD-ROM, identified as *Open Issues List.doc.*

AUDIT

Status Reviews

Reviews should occur throughout the life cycle as identified in the software project management plan. These reviews can be conducted as internal management or customer management reviews. The purpose of these types of reviews is to determine the current status and risk of the software effort. The data presented during these reviews should be used as the basis for decision making regarding the future path and progress of the project.

Internal Management Reviews

The following are suggested agenda items for discussion during an internal management review:

    An overview of current work to date
    Status versus project plan comparison (cost and schedule)
    Discussion of any suggested alternatives (optional)
    Risk review and evaluation
    Updates to the project management plan (if applicable)

Customer Management Reviews

The following are suggested agenda items for discussion during a customer management review:

    An overview of current work to date
    Status versus project plan comparison (cost and schedule)
    Discussion of any suggested alternatives (optional)
    Risk review and evaluation
    Customer approval of any updates to the project plan

Open Issues List
As of: [Date]

Identification

This document contains open issues from the _____ meeting:

Package	
Author	
Review date	
List version	

Open issues

The matrix below lists any *open issues*, and the response to those issues. The disposition column indicates the resolution of the issue.

Ref	Severity	Comment	Source	Response	Disposition

Figure C-16. Open issues list.

Critical Dependencies Tracking

IEEE Std 1490, IEEE Adoption of PMI Standard, A Guide to the Project Management Body of Knowledge, describes the identification, tracking, and control of all items critical to the successful management of a project. The PMI refers to the identification and tracking of critical dependencies as integrated change control. According to the IEEE Std 1490:

> Integrated change control is concerned with a) influencing the factors that create changes to ensure that changes are agreed upon, b) determining that a change has occurred, and c) managing the actual changes when and as they occur. The original defined project scope and the integrated performance baseline must be maintained by continuously managing changes to the baseline, either by rejecting new changes or by approving changes and incorporating them into a revised project baseline. [34]

As described by IEEE Std 1490, there are three key inputs that support integrated change control. The content of the software project management plan provides key input and becomes the change control baseline. Status reporting provides key input, providing interim information on project performance. Change requests also provide a record of the changes requested to items critical to the success of the project.

The work products that support the tracking of critical dependencies are directly related to the three key inputs described above. All changes in the status of critical dependencies affect the project plan and must be reflected in the plan as plan updates. Corrective action, and the documentation of lessons learned, may be triggered as a result of status reporting. Any corrective action, and lessons learned, may be recorded as the resolution of status report action items. Requested changes may be recorded any number of ways—as notes from a status review or, more formally, as part of a change request process.

For integrated project management with other impacting projects, several planning and management tasks must be enhanced for interproject communications. Table C-25 details tasks specifically for IPM SP 2.2—Manage Dependencies.

List of Measures for Reliable Software [7]

Table C-26 provides a list of suggested measures in support of the production of reliable software is provided. This list is a reproduction of content found in IEEE Std 982.1, IEEE Standard Dictionary of Measures to Produce Reliable Software.

Example Measures

Management Category

Manpower Measure. Manpower measures are used primarily for project management and do not necessarily have a direct relationship with other technical

Table C-25. Tasks in support of integrated project management

Critical Dependencies Tracking—Detailed Project Planning
 1. Detail Project Definition and Work Plan Narrative and dependences on Other Projects
 2. Define Work Breakdown Structure (WBS) and dependence on Other Project Plan's WBS
 3. Develop List of Relevant Stakeholders, including Other Projects Stakeholders
 4. Develop Communication Plan (Internal, with Other Projects, and External)
 5. Layout Task Relationships (proper sequence: dependencies, predecessor and successor tasks)
 6. Identify **Critical Dependencies**
 7. Estimate Time lines
 8. Determine the Planned Schedule and Identify Milestones
 9. Determine the Planned Schedule and Identify common Milestones with Other Projects
10. Obtain Stakeholder (including Other Projects) Buyoff and Signoff
11. Obtain Executive Approval to Continue

Critical Dependencies Tracking—Detailed Project Plan Execution
12. Mark Milestones (significant event markers) and **Critical Dependencies**
13. Assign and Level Resources Against Shared Resources Pool
14. Set Project Baseline
15. Set up Issues Log
16. Set up Change Request Log
17. Set up Change of Project Scope Log
18. Record Actual Resource Usage and Costs, including Other Projects
19. Track and Manage Critical Path and **Critical Dependencies**
20. Manage and Report Status as Determined by Project Plan
21. Communicate with Stakeholders, including Other Projects, as determined by Project Plan
22. Report Monthly Project Status, including Other Projects
23. Close Project by Formal Acceptance of All Stakeholders
24. Report on Project Actuals versus Baseline Estimates
25. Complete Post Implementation Review Report—Lessons Learned
26. Complete and File All Project History Documentation in Document Repository
27. Celebrate

and maturity measures. This measure should be used in conjunction with the development progress measure. The value of this measure is somewhat tied to the accuracy of the project plan, as well as to the accuracy of the labor reporting.

This measure provides an indication of the application of human resources to the development program and the ability to maintain sufficient staffing to complete the project. It can also provide indications of possible problems with meeting schedule and budget. It is used to examine the various elements involved in staffing a software project. These elements include the planned level of effort, the actual level of effort, and the losses in the software staff measured per labor category. Planned manpower profiles can be derived from the associated project planning documents.

Table C-26. List of suggested measures for reliable software

Paragraph	Description of Measure
4.1	Fault Density
4.2	Defect Density
4.3	Cumulative Failure Profile
4.4	Fault Days, Number
4.5	Functional or Modular Test Coverage
4.6	Cause and Effect Graphics
4.7	Requirements Traceability
4.8	Defect Indices
4.9	Error Distribution
4.10	Software Maturity Index
4.11	Man Hours per Major Defect Detected
4.12	Number of Conflicting Requirements
4.13	Number of Entries and Exits per Module
4.14	Software Science Measures
4.15	Graph-Theoretic Complexity for Architecture
4.16	Cyclomatic Complexity
4.17	Minimal Unit Test Case Determination
4.18	Run Reliability
4.19	Design Structure
4.20	Mean Time to Discover the Next K Faults
4.21	Software Purity Level
4.22	Estimated Number of Faults Remaining
4.23	Requirement Compliance
4.24	Test Coverage
4.25	Data or Information Flow Complexity
4.26	Reliability Growth Function
4.27	Residual Fault Count
4.28	Failure Analysis Using Elapsed Time
4.29	Testing Sufficiency
4.30	Mean Time to Failure
4.31	Failure Rate
4.32	Software Documentation and Source Listings
4.33	Required Software Reliability
4.34	Software Release Readiness
4.35	Completeness
4.36	Test Accuracy
4.37	System Performance Reliability
4.38	Independent Process Reliability
4.39	Combined Hardware and Software Operational Availability

The planned level of effort is the number of labor hours estimated to be worked on a software module during each tasking cycle. The planned levels are monitored to ensure that the project is meeting the necessary staffing criteria.

Life Cycle Application. The shape of the staff profile trend curve tends to start at a moderate level at the beginning of the contract, grow through design, peak at coding/testing, and diminish near the completion of integration testing. Individual labor categories, however, are likely to peak at different points in the life cycle. The mature result would show little deviation between the planned and actual levels for the entire length of development and scheduled maintenance life cycles.

Algorithm/Graphical Display. The measure graphical display should reflect, by project deliverable software item, the actual versus planned labor hours of effort per month. This measure can be further refined by breaking labor hours into categories (i.e., experienced, novice, or special).

Significant deviations (those greater than 10%) of actual from planned levels indicate potential problems with staffing. Deviations between actual and planned levels can be detected, analyzed, and corrected before they negatively impact the development schedule. Losses in staff can be monitored to detect a growing trend or significant loss of experienced staff. This indicator assists in determining whether there are a sufficient number of employees to produce the product in the tasking cycle.

Data Requirements. Software tasking documentation must reflect the expected staffing profiles and labor hour allocations.

For each labor category tracked, record the:

Labor category name:
a. Experienced
   Programmer:
   Lead:
b. Senior
   Support:

For each experience level per tasking cycle, record the:

Number of planned personnel staffing

Number of personnel actually on staff in current reporting period

Number of unplanned labor hour losses

Number of labor hours that are planned to be expended in next reporting period
   (cumulative)

Number of labor hours that are actually expended in the current reporting period
   (cumulative)

Frequency and Type of Reporting. There will be no less than one report monthly. Reporting by [Project Abbreviation] Program Manager to [Company

Name] corporate management will be in the form of Program Manager's Review (PMR).

Use/Interpretation. Tracking by individual labor category may be done for projects in order to monitor aspects of a particular program that are deemed worthy of special attention. Total personnel are the sum of experienced and support personnel. Special-skills personnel are counted within the broad categories of experienced and support, and may be tracked separately.

Special Skills Personnel. Special skills personnel are defined as those individuals who possess specialized software-related abilities defined as crucial to the success of the particular system. For example, Ada programmers are defined as having skills necessary for completing one type of software project but not others.

Experienced Personnel. Experienced personnel are defined as degreed individuals with a minimum of three years experience in software development for similar applications.

Support Personnel. Support personnel are degreed and nondegreed individuals with a minimum of three years experience in software development, other than those categorized as experienced software engineers.

Reporting Examples. See Figures C-17 and C-18.

Development Progress Measures. Progress measures provide indications of the degree of completeness of the software development effort and can be used to judge readiness to proceed to the next stage of software development. In certain instances, consideration must be given to a possible rebaselining of the software or the addition of modules due to changing requirements.

This measure is used to track the ability to keep Computer Software Unit (CSU) design, code, test, and integration activities on schedule. The development progress measures should be used with the manpower measures to identify specific projects that may be having problems.

These measures should also be used with the breadth, depth, and fault-profile testing measures to assess the readiness to proceed to a formal government test.

Life Cycle Application. Collection is to begin at software requirements review (SRR) and continue for the entirety of the software development. The progress demonstrated in design, coding, unit testing, and integration of CSUs should occur at a reasonable rate. The actual progress shown in these areas versus the originally planned progress can indicate potential problems with a project's schedule.

Algorithm/Graphical Display. The progress measure graphical display should reflect, by project software deliverable item, the percent of planned and actual CSUs

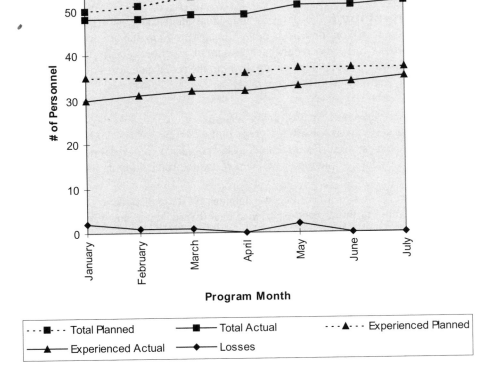

Figure C-17. Manpower staffing profile.

per tasking cycle milestone (for example, percent complete/per month). Completed (actual) CSUs may be subcategorized into percent of CSUs 100% designed, percent of CSUs coded and successfully unit tested, and percent of CSUs 100% integrated.

Additionally, using the requirements traceability matrix, the developed and verified functionality versus time may be plotted as a measure of development progress.

Data Requirements. For project development, test, and integration schedules. For each project, determine the:

1. Number of CSUs
2. Number of CSUs 100% designed
3. Number of CSUs 100% coded and successfully validated
4. Number of CSUs 100% integrated
5. Number of planned CSUs to be 100% designed and reviewed for development cycle

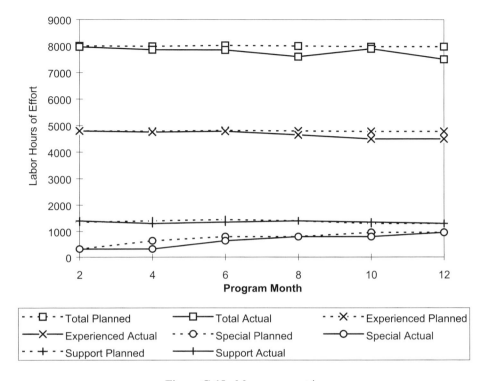

Figure C-18. Manpower metric.

Also include the number of planned CSUs to be 100% coded and successfully validated for the development cycle and the number of planned CSUs to be 100% integrated for the development cycle.

Frequency and Type of Reporting. The [Project Abbreviation] Project Manager shall report progress measure information to the [Project Abbreviation] Program Manager no less than once a month. Also, there shall be no less than one report monthly by the Program Manager to [Company Name] corporate management and the customer in the form of a Program Manager's Review (PMR).

Use/Interpretation. It is important to remember that these measures pass no judgment on whether the objectives in the development plan can be achieved. Special attention should be paid to the development progress of highly complex CSUs.

Reporting Example. See Figure C-19.

Schedule Measures. These measures indicate changes and adherence to the planned schedules for major milestones, activities, and key software deliverables. Software activities and delivery items may be tracked using the schedule measures.

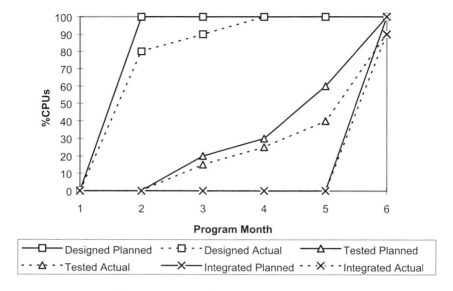

Figure C-19. Development progress measure.

The schedule measures may be used with other measures to help judge program risk. For example, it could be used with the test coverage measures to determine if there is enough time remaining on the current schedule to allow for the completion of all testing.

Life Cycle Application. The Project Manager will begin data collection at project start and continue for the entire software development cycle.

Algorithm/Graphical Display. The Project Manager will plot planned and actual schedules for major milestones and key software deliverables as they change over the tasking cycle. Any milestone event of interest may be plotted.

The schedule measures may be plotted as they change over time (reflecting milestone movement), providing indications of problems in meeting key events or deliveries. The further to the right of the trend line for each event, the more problems are being encountered.

Data Requirements. All projects require the following:

1. Software project plan
2. Major Milestone schedules
3. Delivery schedule for key software items

For the activity or event tracked during tasking period, the following are required:

1. Event name
2. Date of report
3. Planned activity start date
4. Actual activity start date
5. Planned activity end date
6. Actual activity end date

Frequency and Type of Reporting. The [Project Abbreviation] Project Manager will report schedule information to the [Project Abbreviation] Program Manager and the customer no less than bi-weekly. This reporting will be accomplished as part of scheduled staff meetings. Also, there shall be no less than one report monthly by the Program Manager to [Company Name] corporate management, and the customer, in the form of Program Manager's Review (PMR).

Use/Interpretation. No formal evaluation criteria for the trend of the schedule measures are given. Large slippages are indicative of problems. Maintaining conformance with calendar-driven schedules should not be used as the basis for proceeding beyond milestones. The schedule measure passes no judgment on the achievability of the development plan.

Reporting Example. See Figure C-20.

Requirements Category

Requirements Traceability Measure. The requirements traceability measure measures the adherence of the software products to their requirements throughout the development life cycle. The technique to performing this employs the development of a project software requirements traceability matrix (SRTM). (For additional information, refer to the [Project Abbreviation] Software Requirements Specification, Appendix A).

The SRTM is the product of a structured, top-down, hierarchical analysis that traces the software requirements through the design to the code and test documentation. The requirements traceability measures should be used in conjunction with the test coverage measures (depth and breadth of testing) and the development progress measure (optional) to verify if sufficient functionality has been demonstrated to warrant proceeding to the next stage of development or testing. They should also be used in conjunction with the design stability and requirements stability measures.

Life Cycle Application. The Project Manager will begin tracing measures during user requirements definition, updating the trace in support of major milestones or at major software release points.

By the nature of the software development process, especially in conjunction with an evolutionary development strategy, the trace of requirements is an iterative process. That is, as new software releases add more functionality to the system, the requirements trace will have to be revisited and augmented.

STATUS

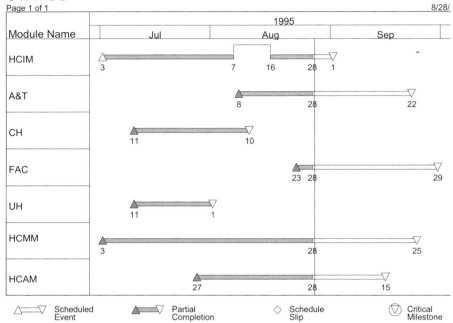

Figure C-20.

Algorithm/Graphical Display. This measure is a series of percentages that can be calculated from the matrix described. These will include:

- Percent baseline software requirements specification requirements in CSU design
- Percent software requirements in code
- Percent software requirements having test cases identified

Tracing from code to requirements should also be accomplished. Each CSU is examined for the requirements it satisfies and the percentage of open or additional requirements it traced. Backward tracing of requirements at the early stages can identify requirements that have been added in the requirements decomposition process. Each requirement added to the specification should be attributable in some way to a higher-level requirement from a predecessor requirement.

Data Requirements. For all projects:

1. Documented/approved software requirements specification (SRS)
2. Documented/approved software design document (SDD)

3. Software test description in accordance with DOD-STD-498
4. Completed SRTM

For each project software deliverable:

1. Number of SRS software requirements:
   - Total traceable to SRS
   - Traceable to SDD
   - Traceable to code
   - Traceable to system test

Frequency and Type of Reporting. The Program Manager will update periodically in support of milestones or major releases. This tracing should be a key tool used at all system requirement and design reviews. It can serve to indicate those areas of requirements or software design that have not been sufficiently thought out. The trend of the SRTM should be monitored over time for closure.

Use/Interpretation. A benefit of requirements traceability is that those modules that appear most often in the matrix (thus representing the ones that are most crucial in that they are required for multiple functions or requirements) can be highlighted for earlier development and increased test scrutiny.

Example. An example requirements traceability matrix follows.

[Project Abbreviation] Requirements Traceability Matrix. (For additional information regarding requirements elicitation refer to [Project Acronym] Software Requirements Management Plan.)

Requirement Name	Priority	Risk	SRS Paragraph	Formal Test Paragraph	Test Activity
Input Personnel Data	H	L	3.1.1	6.3.14	D
Store Personnel Data	M	H	3.1.2	6.3.17	T, A

Verification Method(s): D = Demonstration, T = Test, A = Analysis, I = Inspection.
   Field Descriptions:

1. Requirement Name	A short description of the requirement to be satisfied
2. Priority	High (H), Medium (M), Low (L). To be negotiated with customer and remains fixed throughout software development life cycle
3. Risk	High (H), Medium (M), Low (L). Determined by technical staff and will change throughout software development life cycle.
4. SRS Paragraph	Paragraph number from Section 3 of the SRS.
5. Validation Method(s)	Method to be used to validate that the requirement has been satisfied.
6. Formal Test Paragraph	This column will provide a linkage between the SRS and the software Test Plan. This will indicate the test to be executed to satisfy the requirement.

Requirements Stability Measure. Requirements stability measures indicate the degree to which changes in the software requirements and/or the misunderstanding of requirements implementation affect the development effort. It also allows for determining the cause of requirements changes. The measures for requirements stability should be used in conjunction with those for requirements traceability, fault profiles, and the optional development progress.

Life Cycle Application. Changes in requirements tracking begin during requirements definition and continue to the conclusion of the development cycle. It is normal that when program development begins, the details of its operation and design be incomplete. It is normal to experience changes in the specifications as the requirements become better defined over time. When design reviews reveal inconsistencies, a discrepancy report is generated. Closure is accomplished by modifying the design or the requirements. When a change is required that increases the scope of the project, a baseline change request (BCR) should be submitted.

Algorithm/Graphical Display. This measure may be presented as a series of percentages representing:

1. The percent of requirements discrepancies, both cumulative and cumulative closed, over the life of the project, and cumulative requirements discrepancies over time, versus closure of those discrepancies. Good requirements stability is indicated by a leveling off of the cumulative discrepancies curve with most discrepancies having reached closure.
2. The percent of requirements changed or added, noncumulative, spanning project development and the effect of these changes in requirements in lines of code. Several versions of this type of chart are possible. One version may show the number of change enhancement requests (CERs) and affected lines of code. Additionally, it is also possible to look at the number of modules affected by requirements changes.

The plot of open discrepancies can be expected to spike upward at each review and to diminish thereafter as the discrepancies are closed. For each engineering change, the amount of software affected should be reported in order to track the degree to which BCRs increase the difficulty of the development effort. Only those BCRs approved by the associated CCB should be counted.

Data Requirements. A "line of code" will differ from language to language, as well as from programming style to programming style. In order to consistently measure source lines of code, count all noncommented, nonblank, executable and data statements (standard noncommented lines of code or SNCLC).

For each project, calculate:

1. Number of software requirements discrepancies as a result of each review (software requirements review, preliminary design review, etc.)

2. Number of baseline change requests (BCRs) generated from changes in requirements
3. Total number of SNCLC
4. Total number of SRS requirements
5. Percentage of SRS requirements added due to approved BCRs
6. Percentage of SRS requirements modified due to approved BCRs
7. Percentage of SRS requirements deleted due to approved BCRs

Frequency and Type of Reporting. The Project Manager will update the measures periodically in support of milestones or major releases. This tracing should be a key tool used at all system requirement and design reviews.

Use/Interpretation. Causes of project turbulence can be investigated by looking at requirements stability and design stability together. If design stability is low and requirements stability is high, the designer/coder interface is suspect. If design stability is high and requirements stability is low, the interface between the user and the design activity is suspect. If both design stability and requirements stability are low, both the interfaces between the design activity and the code activity and between the user and the design activity are suspect.

Allowances should be made for higher instability in the case in which rapid prototyping is utilized. At some point in the development effort, the requirements should be firm so that only design and implementation issues will cause further changes to the specification.

A high level of instability at the critical design review (CDR) stage indicates serious problems that must be addressed prior to proceeding to coding.

Example. See Figure C-21.

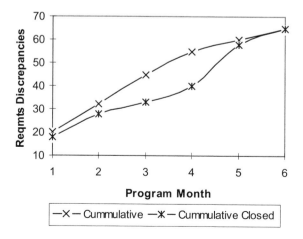

Figure C-21. Requirements stability example.

Quality Category

Design Stability Measure. Design stability is used to indicate the quantity of change made to a design. The design progress ratio shows how the completeness of a design advances over time and helps give an indication of how to view the stability in relation to the total projected design. The design stability measure can be used in conjunction with the complexity measure to highlight changes to the most complex modules. It can also be used with the requirements measures to highlight changes to modules that support the most critical user requirements. Design stability does not assess the quality of design. Other measures (e.g., complexity) can contribute to such an evaluation.

Life Cycle Application. Begin tracking no later than preliminary design review (PDR) and continue for each version until completion.

Algorithm/Graphical Display. Items to be measured:

M = Number of CSUs in current delivery/design

F_c = Number of CSUs in current delivery/design that include design

F_a = Number of CSUs in current delivery/design that are additions to previous delivery

F_d = Number of CSUs in previous delivery/design that have been deleted

T = Total modules projected for project

Formulae:

$$S \text{ (stability)} = [M - (F_a + F_c + F_d)]/M$$

Note: It is possible for stability to be a negative value. This may indicate that everything previously delivered has been changed and that more modules have been added or deleted.

$$DP \text{ (design progress ratio)} = M/T$$

Note: If some modules in the current delivery are to be deleted from the final delivery, it is possible for design progress to be greater than one.

Data Requirements. For each project and version, calculate:

1. Date of completion
2. M
3. F_c
4. F_a
5. F_d

6. *T*

7. *S*

8. *DP*

Frequency and Types of Reporting. The Project Manager will update periodically in support of milestones or at major releases. Design stability should be monitored to determine the number and potential impact of design changes, additions, and deletions on the software configuration.

Use/Interpretation. The trend of design stability provides an indication of whether the software design is approaching a stable state, or leveling off, of the curve at a value close to or equal to one. It is important to remember that during periods of inactivity, the magnitude of the change is relatively small or diminishing and may be mistaken for stability.

In addition to a high value and level curve, the following other characteristics of the software should be exhibited:

a) The development progress measure is high.

b) Requirements stability is high.

c) Depth of testing is high.

d) The fault profile curve has leveled off and most software trouble reports [test problem reports (TPRs) or internal test reports (ITRs)] have been fixed or voided.

This measure does not measure the extent or magnitude of the change within a module or assess the quality of the design. Other measures (e.g., complexity) can contribute to this evaluation.

Allowances should be made for lower stability in the case of rapid prototyping or other development techniques that do not follow the [Project Abbreviation] life cycle model for software development (for additional information refer to the [Project Abbreviation] Software Quality Assurance Plan). An upward trend with a high value for both stability and design progress is recommended before release.

Experiences with similar projects should be used as a basis for comparison. Over time, potential thresholds may be developed for similar types of projects.

Reporting Example. See Figure C-22.

Breadth of Testing Measure. Breadth of testing addresses the degree to which required functionality has been successfully demonstrated as well as the amount of testing that has been performed. This testing can be called "black box" testing, since one is only concerned with obtaining correct outputs as a result of prescribed inputs.

The organization should clearly specify what functionality should be in place at each milestone during the development life cycle. At each stage of testing (unit

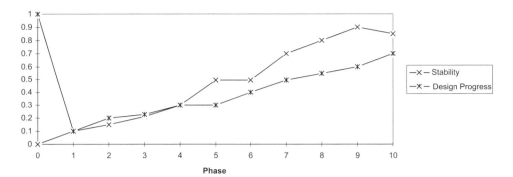

Figure C-22. Example design stability/progress measure.

through system), emphasis should be placed on demonstrating that a high percentage of the functionality needed for that stage of testing is achieved.

Life Cycle Application. Begin collecting at the end of unit testing. Each priority level of the requirements test case to be used for evaluating the success of a requirement should be developed in coordination with project test. This is to ensure that sufficient test cases are generated to adequately demonstrate the requirements.

Algorithm/Graphical Display. Breadth of testing consists of three different measures. One measure deals with coverage and two measures deal with success. These three subelements are portrayed in the following equation:

$$\underbrace{\frac{\#\,\text{CSUs tested}}{\text{total}\,\#\,\text{CSUs}}}_{Coverage} \times \underbrace{\frac{\#\,\text{CSUs passed}}{\#\text{CSUs tested}}}_{Test\ Success} = \underbrace{\frac{\#\,\text{CSUs passed}}{\text{total}\,\#\,\text{CSUs}}}_{Overall\ Success}$$

Breadth of Testing Coverage. Breadth of testing coverage is computed by dividing the number of requirements that have been tested (with all applicable test cases under both representative and maximum stress loads) by the total number of requirements.

Breadth of Testing Success. Breadth of testing success is computed by dividing the number of requirements that have been successfully demonstrated through testing by the number of requirements that have been tested.

Breadth of Testing Overall Success. Breadth of testing overall success is computed by dividing the number of requirements that have been successfully demonstrated through testing by the total number of requirements.

Data Requirements. For each project, indicate:

1. The number of SRS requirements and associated priorities
2. Number of SRS requirements tested with all planned test cases
3. Number of SRS requirements and associated priorities successfully demonstrated through testing
4. For each of the four priority levels for additional requirements (SPR, BCR):
   a) Number of additional requirements tested with all planned test cases
   b) Number of additional requirements demonstrated through testing
5. For each of the four Alpha/Beta (ITR, TPR) test requirement priority levels:
   a) Number of requirements
   b) Number of requirements validated with planned test cases
   c) Number of requirements successfully demonstrated through testing
6. Test Identification (e.g., unit test, alpha test, beta test)

Frequency and Types of Reporting. All three measures of breadth of testing will be tracked for all project requirements. The results of each measure will be reported at each project level throughout software functional and system level testing at each project milestone and to [Company Name] corporate management monthly.

It is suggested that coverage or success values be expressed as percentages by multiplying each value by 100 before display, in order to facilitate understanding and commonality among measures presentations.

Use/Interpretation. One of the most important measures reflected by this measure is the categorization of requirements in terms of priority levels. With this approach, the most important requirements can be highlighted. Using this prioritization scheme, one can partition the breadth of testing measure to address each priority level. At various points along the development path, the pivotal requirements for that activity can be addressed in terms of tracing, test coverage, and test success.

Breadth of Testing Coverage. The breadth of testing coverage indicates the amount of testing performed without regard to success. By observing the trend of coverage over time, it is possible to derive the full extent of testing that has been performed.

Breadth of Testing Success. The breadth of testing success provides indications about requirements that have been successfully demonstrated. By observing the trend of the overall success portion of breadth of testing over time, one gets an idea of the growth in successfully demonstrated functionality.

Failing one test case results in a requirement not being fully satisfied. If sufficient resources exist, breadth of testing may be addressed by examining each requirement in terms of the percent of test cases that have been performed and passed. In this way, partial credit for testing a requirement can be shown (assuming multiple test cases exist for a requirement), as opposed to an "all or nothing" approach. This method, which is not mandated, may be useful in providing additional granularity in breadth of testing.

Example. See Figure C-23.

Depth of Testing Measure. The depth of testing measure provides indications of the extent and success of testing from the point of view of coverage of possible paths and conditions within the software. Depth of testing consists of three separate measures, each of which is comprised of one coverage and two success subelements (similar to breadth of testing). The testing can be called "white box" testing, since there is visibility into the paths and conditions within the software.

This measure should be used in conjunction with requirements traceability and fault profiles and the optional complexity and development progress measures. They must also be used with breadth of testing measures to insure that all aspects of testing are consistent with beta test requirements.

Life Cycle Application. Begin collecting data at critical design review (CDR) and continue through development as changes occur in design, implementation, or testing. Revisit as necessary.

Algorithm/Graphical Display. Graphically, this measure is represented as a composite of the module-level path measures. This should reflect test coverage and overall success of testing.

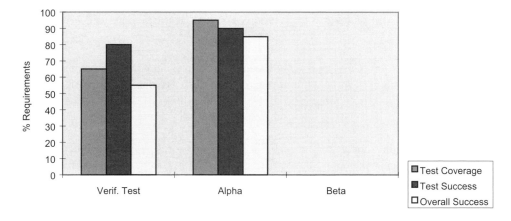

Figure C-23. Breadth of testing example.

Module-Level Measures.

a) Path measure. Defined as the number of paths* in the module that have been successfully executed at least once, divided by the total number of paths in the module.

b) Statement measure. The number of executable statements in each module that have been successfully exercised at least once, divided by the total number of executable statements in the module.

c) Domain measure. The number of input instances that have been successfully tested with at least one legal entry and one illegal entry in every field of every input parameter, divided by the total number of input instances in the module.

d) Decision point measure (optional). The number of decision points in the module that have been successfully exercised with all classes of legal conditions as well as one illegal condition (if any exist) at least once, divided by the total number of decision points in the module. Each decision point containing an "or" should be tested at least once for each of the condition's logical predicates.

Data Requirements. For each module, include:

1. Software project name
2. Module identification
3. Number of paths
4. Number of statements
5. Number of input instances
6. Number of paths tested
7. Number of statements tested
8. Number of input instances tested
9. Number of decision points tested
10. Number of paths successfully executed (at least once)
11. Number of statements that have been successfully exercised
12. Number of inputs successfully tested with both legal and illegal entries
13. Number of decision points (optional)
14. Number of decision points that have been successfully exercised at least once with all legal classes of conditions and one illegal condition (optional)

Frequency and Type of Reporting. Report only those modules that have been modified or further tested after depth of testing values have been reported for the first time. In recognition of the effort required to collect and report this measure, the following rules are offered:

---

*A path is defined as a logical traversal of a module, from an entry point to an exit point.

1. Always report the domain measure.
2. Compute the path and statement measure if provided with automated tools.

Use/Interpretation. The depth of testing measures addresses the issues of test coverage, test success, and overall success by considering the paths, statements, inputs and decision points of a software module.

These measures are to be initially collected at the module level, but this may be extended to the system level if appropriate. Early in the testing process, it makes sense to assess depth of testing at the module level, but later it may make more sense to consider the system in its entirety. This determination is made by the [Project Abbreviation] Program Manager in conjunction with System Test/Validation.

Reporting Example. See Figure C-24.

Fault Profiles Measure. Fault profiles provide insight into the quality of the software, as well as the contractor's ability to fix known faults. These insights come from measuring the lack of quality (i.e., faults) in the software.
Life Cycle Application. Begin after completion of unit testing when software has been brought under configuration control and continue through software maintenance.

Algorithm/Graphical Display. For each software project plot:

a) The cumulative number of detected and closed software faults (i.e., TPR/ITRs) will be tracked as a function of time. If manpower allows, one plot will be developed for each priority level.
b) The number of software faults detected and software faults closed tracked as a function of time.

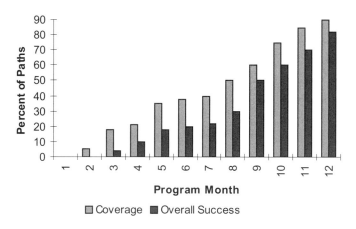

Figure C-24. Depth of testing example.

c) Open faults, by module and priority.
d) Calculate the average age of closed faults as follows. For all closed TPR/ITRs, sum the days from the time the STR was opened and when it was closed. Divide this by the total number of closed problem reports. If manpower allows, this should be calculated for each problem priority or overall.

Data Requirements. The following information can be derived from all change enhancement requests (CERs) that are of the TPR, SPR, or ITR variety:

1. CER unique identifier
2. Date initiated
3. Descriptive title of problem
4. Detailed description of problem
5. Priority:
   a) 1 = causes mission-essential function to be disabled or jeopardizes personnel safety. No work-around.
   b) 2 = causes mission essential-function to be degraded. There is a reasonable work-around.
   c) 3 = causes operator inconvenience but does not affect a mission-essential function.
   d) 0 = all other errors.
6. Category:
   a) Requirements = baseline change request (BCR)
   b) Alpha test = internal test report (ITR)
   c) Beta test = test problem report (TPR)
   d) User-driven changes to data = software problem report (SPR)
   e) User-driven change to interfaces = interface change notice (ICN)
7. Status:
   a) Open
   b) Voided
   c) Hold
   d) Testing (alpha)
   e) Fixed
8. Date detected
9. Date closed
10. Software project, module and version
11. Estimated required effort and actual effort required
12. The following reports may be generated using the SCRTS reporting capabilities. These items are the building block for graphical representations and fault profiles. Indicate (for each module by priority):

   a) Cumulative number of CERs
   b) Cumulative number of closed CERs
   c) Average age of closed CERs
   d) Average age of open CERs
   e) Average age of CERs (both open and closed)
   f) Totals for each CER category (described above)

Frequency and Type of Reporting. The Project Manager will update periodically in support of milestones or at major releases. Design stability should be monitored to determine the number and potential impact of design changes, additions, and deletions on the software configuration. Types of items examined:

1. Displays of detected faults versus closed (verified) faults should be examined for each problem's priority level and for each module. Applied during the early stages of development, fault profiles measure the quality of the translation of the software requirements into the design. CERs opened during this time may suggest that requirements are either not being defined or interpreted correctly. Applied later in the development process, assuming adequate testing, fault profiles measure the implementation of requirements and design into code. CERs opened during this stage could be the result of having an inadequate design to implement those requirements, or a poor implementation of the design into code. An examination of the fault category should provide indications of these causal relationships.

2. If the cumulative number of closed ITR/TPRs remains constant over time and a number of them remain open, this may indicate a lack of problem resolution. The age of the open STRs should be checked to see if they have been open for an unreasonable period of time. If so, these areas need to be reported to the [Project Abbreviation] Program Manager as areas of increased risk.

3. The monthly noncumulative ITR totals for each project can be compared to the TPR totals to provide insights into the adequacy of the alpha validation program.

4. Open-age histograms can be used to indicate which modules are the most troublesome with respect to fixing faults. This may serve to indicate that the project team may need assistance.

5. Average-age graphs can track whether the time to close faults is increasing with time, which may be an indication that the developer is becoming saturated or that some faults are exceedingly difficult to fix. Special consideration should be paid to the gap between open and fixed CERs. If a constant gap or a continuing divergence is observed, especially as a major milestone is approached, the Program Manager should take appropriate action. Once an average CER age has been established, large individual deviations should be investigated.

Use/Interpretation. Early in the development process, fault profiles can be used to measure the quality of the translation of the software requirements into the design. Later in the process, they can be used to measure the quality of the implementation of the software requirements and design into code.

If fault profile tracking starts early in software development, an average CER open age of less than three months may be experienced. After release, this value may be expected to rise.

Reporting Examples. Cumulative by month, see Figure C-25. By month, see Figure C-26. Average age/open faults, see Figure C-27.

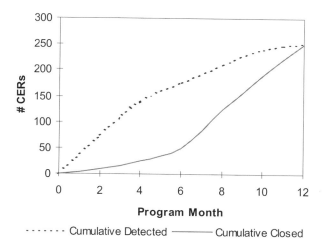

Figure C-25. Example of cumulative by month reporting.

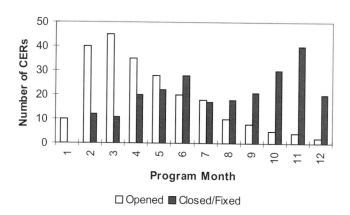

Figure C-26. Example of project CERs by month reporting.

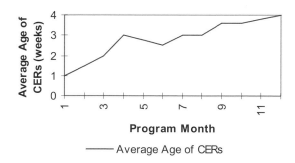

Figure C-27. Example of average age of CERs by project reporting.

Measurement Information Model in ISO/IEC 15939

Table C-27 provides three examples of instantiations of the Measurement Information Model as presented in ISO/IEC 15939 Software Engineering—Software Measurement Process [56].

PROBLEM RESOLUTION

Risk Taxonomy

The taxonomy in Table C-28 was developed by the SEI [66] in support of risk management process development.

Risk Taxonomy Questionnaire

The questionnaire in Table C-29 was developed by the SEI [66] in support of risk management process development. This questionnaire can be extremely helpful when trying to identify and categorize project risk.

Risk Action Request

Table C-30 provides the suggested content for a risk action request that may be used to effectively support the assessment of project risk. The recommended content shown are based on the requirements found in IEEE Std 1540-2001, IEEE Standard for Software Lifecyle Processes—Risk Management, Annex B: Risk Action Request.

Risk Mitigation Plan

Table C-31 provides suggested content for a risk mitigation plan that may be used to document and control project risk. The recommended content shown is based on

Table C-27. Measurement information model in ISO/IEC 15939.

	Productivity	Quality	Progress
Information Need	Estimate productivity of future project	Evaluate product quality during design	Assess status of coding activity
Measurable Concept	Project productivity	Product quality	Activity status
Relevant Entities	1. Code produced by past projects 2. Effort expended by past projects	1. Design packages 2. Design inspection reports	1. Plan/schedule 2. Code units completed or in progress
Attributes	1. C++ language statements (in code) 2. Time card entries (recording effort)	1. Text of inspection packages 2. Lists of defects found in inspections	1. Code units identified in plan 2. Code unit status
Base Measures	1. Project X lines of code 2. Project X hours of effort	1. Package X size 2. Total defects for package X	1. Code units planned to date 2. Code unit status
Measurement Method	1. Count semicolons in Project X code 2. Add time card entries together for Project X	1. Count number of lines of text for each package 2. Count number of defects listed in each report	1. Count number of code units scheduled to be completed by this date 2. Ask programmer for percent complete of each unit
Type of Measurement Method	1. Objective 2. Objective	1. Objective 2. Objective	1. Objective 2. Subjective
Scale	1. Integers from zero to infinity 2. Real numbers from zero to infinity	1. Integers from zero to infinity 2. Integers from zero to infinity	1. Integers from zero to infinity 2. Integers from zero to one hundred
Type of Scale	1. Ratio 2. Ratio	1. Ratio 2. Ratio	1. Ratio 2. Ordinal
Unit of Measurement	1. Line 2. Hour	1. Lines 2. Defects	1. Code unit 2. One-hundredth code unit
Derived Measure	Project X productivity	Inspection defect density	Progress to date
Measurement Function	Divide Project X lines of code by Project X hours of effort	Divide total defects by package size for each package	Add status for all code units planned to be complete to date
Indicator	Average productivity	Design defect density	Code status expressed as a ratio

(continued)

Table C-27. *Continued*

	Productivity	Quality	Progress
Model	Compute mean and standard deviation of all project productivity values	Compute process center and control limits using values of defect density	Divide progress to date by (code units planned to date times 100)
Decision Criteria	Computed confidence limits based on the standard deviation indicate the likelihood that an actual result close to the average productivity will be achieved. Very wide confidence limits suggest a potentially large departure and the need for contingency planning to deal with this outcome.	Results outside the control limits require further investigations.	Resulting ratio should fall between 0.9 and 1.1 to conclude the project is on schedule.

the requirements found in IEEE Std 1540, IEEE Standard for Software Lifecyle Processes—Risk Management Annex C: Risk Treatment Plan.

Risk Matrix Sample

A risk matrix is a helpful tool when used to support risk analysis activities [71]. Table C-32 provides an example of a risk matrix. Tables C-33 and Table C-34 provide description of the associated severity and probability levels.

MANAGEMENT

Work Breakdown Structure

A work breakdown structure (WBS) defines and breaks down the work associated with a project into manageable parts. It describes all activities that have to occur to accomplish the project. The WBS serves as the foundation for the development of project schedules, budget, and resource requirements.

A WBS may be structured by project activities or components, functional areas or types of work, or types of resources, and is organized by its smallest component—a work package. A work package is defined as a deliverable or product at the lowest level of the WBS. Work packages may also be further subdivided into activities or tasks. IEEE Std 1490, IEEE Guide—Adoption of the PMI Standard, A Guide to the Project Management Body of Knowledge, recommends the use of

Table C-28. Risk taxonomy

A. Product Engineering	B. Development Environment	C. Program Constraints
1. Requirements	1. Development Process	1. Resources
a. Stability	a. Formality	a. Schedule
b. Completeness	b. Suitability	b. Staff
c. Clarity	c. Process Control	c. Budget
d. Validity	d. Familiarity	d. Facilities
e. Feasibility	e. Product Control	2. Contract
f. Precedent	2. Development System	a. Type of contract
g. Scale	a. Capacity	b. Restrictions
2. Design	b. Suitability	c. Dependencies
a. Functionality	c. Usability	3. Program Interfaces
b. Difficulty	d. Familiarity	a. Customer
c. Interfaces	e. Reliability	b. Associate
d. Performance	f. System Support	Contractors
e. Testability	g. Deliverability	c. Subcontractor
f. Hardware	3. Management Process	d. Prime Contractor
Constraints	a. Planning	e. Corporate
g. Nondevelopmental	b. Project Organization	Management
Software	c. Management	f. Vendors
3. Code and Unit Test	Experience	g. Politics
a. Feasibility	d. Program Interfaces	
a. Testing	4. Management Methods	
c. Coding/Implementa-	a. Monitoring	
tion	b. Personnel	
4. Integration and Test	Management	
a. Environment	c. Quality Assurance	
b. Product	d. Configuration	
c. System	Management	
5. Engineering	5. Work Environment	
Specialties	a. Quality Attitude	
a. Maintainability	b. Cooperation	
b. Reliability	c. Communication	
c. Safety	d. Morale	
d. Security		
e. Human Factors		
f. Specifications		

nouns to represent the "things" in a WBS. Figure C-28 provides an example of a sample WBS organized by activity.

Work Flow Diagram

Work flow may be defined as the set of all activities in a project from start to finish. A work flow diagram is a pictorial representation of the operational aspect of a work procedure: how tasks are structured, who performs them, what their relative

Table C-29. Risk taxonomy questionnaire

A. Product Engineering. Technical aspects of the work to be accomplished.
  1. Requirements
    a. Stability. Are requirements changing even as the product is being produced?
    b. Completeness. Are requirements missing or incompletely specified?
    c. Clarity. Are the requirements unclear or in need of interpretation?
    d. Validity. Will the requirements lead to the product the customer has in mind?
    e. Feasibility. Are there requirements that are technically difficult to implement?
    f. Precedent. Do requirements specify something never done before or beyond the experience of program personnel?
    g. Scale. Is the system size or complexity a concern?
  2. Design
    a. Functionality. Are there any potential problems in designing to meet functional requirements?
    b. Difficulty. Will the design and/or implementation be difficult to achieve?
    c. Interfaces. Are internal interfaces (hardware and software) well defined and controlled?
    d. Performances. Are there stringent response time or throughput requirements?
    e. Testability. Is the product difficult or impossible to test?
    f. Hardware Constraints. Does the hardware limit the ability to meet any requirements?
    g. Nondevelopmental Software. Are there problems with software used in the program but not developed by the program?
  3. Code and Unit Test
    a. Feasibility. Is the implementation of the design difficult or impossible?
    b. Testing. Is the specified level and time for unit testing adequate?
    c. Coding/Implementation. Are the design specifications in sufficient detail to write code? Will the design be changing while coding is being done?
  4. Integration and Test
    a. Environment. Is the integration and test environment adequate? Are there problems developing realistic scenarios and test data to demonstrate any requirements?
    b. Product. Is the interface definition inadequate, facilities inadequate, or time insufficient? Are there requirements that will be difficult to test?
    c. System. Has adequate time been allocated for system integration and test? Is system integration uncoordinated? Are interface definitions or test facilities inadequate?
  5. Engineering Specialties
    a. Maintainability. Will the implementation be difficult to understand or maintain?
    b. Reliability. Are reliability or availability requirements allocated to the software? Will they be difficult to meet?
    c. Safety. Are the safety requirements infeasible and not demonstrable?
    d. Security. Are there unprecedented security requirements?
    e. Human Factors. Is there any difficulty in meeting the human factor requirements?
    f. Specifications. Is the documentation adequate to design, implement, and test the system?

Table C-29. *Continued*

B. Development Environment. Methods, procedures, and tools in the production of the software products.
1. Development Process
   a. Formality. Will the implementation be difficult to understand or maintain?
   b. Suitability. Is the process suited to the development mode, for example, spiral, prototyping? Is the development process supported by a compatible set of procedures, methods, and tools?
   c. Process Control. Is the software development process enforced, monitored, and controlled using measures?
   d. Familiarity. Are the project members experienced in use of the process? Is the process understood by all project members?
   e. Product Control. Are there mechanisms for controlling changes in the product?
2. Development System
   a. Capacity. Are there enough workstations and processing capacity for all the staff?
   b. Suitability. Does the development system support all phases, activities, functions of the program?
   c. Usability. Do project personnel find the development system easy to use?
   d. Familiarity. Have project personnel used the development system before?
   e. Reliability. Is the system considered reliable?
   f. System Support. Is there timely expert or vendor support for the system?
   g. Deliverability. Are the definition and acceptance requirements defined for delivering the system to the customer?
3. Management Process
   a. Planning. Is the program managed according to a plan?
   b. Project Organization. Is the program organized effectively? Are the roles and reporting relationships well defined?
   c. Management Experience. Are the managers experienced in software development, software management, the application domain, and the development process?
   d. Program Interfaces. Is there a good interface with the customer and is the customer involved in decisions regarding functionality and operation?
4. Management Method
   a. Monitoring. Are management measures defined and is development progress tracked?
   b. Personnel Management. Are project personnel trained and used appropriately?
   c. Quality Assurance. Are there adequate procedures and resources to assure product quality?
   d. Configuration Management. Are the change procedures or version control, including installation site(s) adequate?
5. Work Environment
   a. Quality Attitude. Does the project lack orientation toward quality work?
   b. Cooperation. Does the project lack team spirit? Does conflict resolution require management intervention?
   c. Communication. Does the project lack awareness of mission, goals, or communication of technical information among peers and managers?
   d. Morale. Is there a nonproductive, noncreative atmosphere? Does the project lack rewards or recognition for superior work?

(*continued*)

Table C-29. *Continued*

C. Program Constraints. Methods, procedures, and tools in the production of the software products.
1. Resources
   a. Schedule. Is the project schedule inadequate or unstable?
   b. Staff. Is the staff inexperienced, lacking domain knowledge, lacking skills, or not inadequately sized?
   c. Budget. Is the funding insufficient or unstable?
   d. Facilities. Are the facilities inadequate for building and delivering the product?
2. Contract
   a. Type of contract. Is the contract type a source of risk to the project?
   b. Restrictions. Does the contract include any inappropriate restrictions?
   c. Dependencies. Does the program have any critical dependencies on outside products or services?
3. Program Interfaces
   a. Customer. Are there any customer problems such as a lengthy document-approval cycle, poor communication, or inadequate domain expertise?
   b. Associate Contractors. Are there any problems with associate contractors such as inadequately defined or unstable interfaces, poor communication, or lack of cooperation?
   c. Subcontractor. Is the program dependent on subcontractors for any critical areas?
   d. Prime Contractor. Is the program facing difficulties with its prime contractor?
   e. Corporate Management. Is there a lack of support or micromanagement from upper management?
   f. Vendors. Are vendors unresponsive to program needs?
   g. Politics. Are politics causing a problem for the program?

Table C-30. Risk action request suggested content

Unique Identifier
Date of Issue and Status
Approval Authority
Scope
Request Originator
Risk Category
Risk Threshold
Project Objectives
Project Assumptions
Project Constraints
Risk Description
Risk Likelihood and Timing
Risk Consequences
Risk Mitigation Alternatives
   Descriptions
   Recommendation
   Justification
Disposition

Table C-31. Risk mitigation plan suggested content

Unique Identifier
Cross-reference to Risk Action Request Unique
    Identifier
Date of Issue and Status
Approval Authority
Scope
Request Originator
Planned Risk Mitigation Activities and Tasks
Performers
Risk Mitigation Schedule (including resource
    allocation)
Mitigation Measures of Effectiveness
Mitigation Cost and Impact
Mitigation Plan Management Procedures

Table C-32. Sample risk matrix

Probability Severity	Frequent	Probable	Occasional	Remote	Improbable
Catastrophic	IN	IN	IN	H	M
Critical	IN	IN	H	M	L
Serious	H	H	M	L	T
Minor	M	M	L	T	T
Negligible	M	L	T	T	T

Legend: T = Tolerable, L = Low, M = Medium, H = High, IN = Intolerable.

Table C-33. Risk severity levels description

Severity	Consequence
Catastrophic	Greater than 6 month slip in schedule; greater than 10% cost overrun; greater than 10% reduction in product functionality
Critical	Less than 6-month slip in schedule; less than 1% cost overrun; less than 10% reduction in product functionality
Serious	Less than 3 month slip in schedule; less than 5% cost overrun; less than 5% reduction in product functionality
Minor	Less than 1 month slip in schedule; less than 2% cost overrun; less than 2% reduction in product functionality

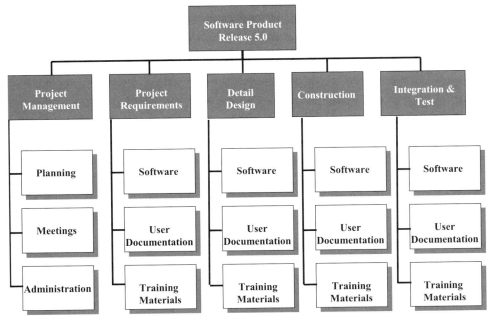

Figure C-28. WBS organized by activity [34].

order is, how they are synchronized, how information flows to support the tasks, and how tasks are being tracked. Work flow problems can be modeled and analyzed using Petri nets. Petri nets are abstract, formal models of information flow, showing static and dynamic properties of a system [2]. Figure C-29 provides an example of a work flow diagram.

Stakeholder Involvement Matrix

The matrix shown in Table C-35 presents a consolidated view of stakeholder involvement. This matrix can be used to define and document stakeholder involvement. Table C-34 provides an example of a stakeholder involvement matrix. An

Table C-34. Risk probability levels description

Probability	Description
Frequent	Anticipate occurrence several times a year (>10 events)
Probable	Anticipate occurrence repeatedly during the year (2 to 10 events)
Occasional	Anticipate occurrence some time during the year (1 event)
Remote	Occurrence is unlikely though conceivable (< 1 event per year)

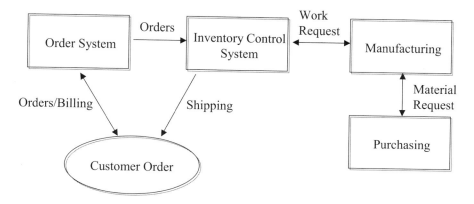

Figure C-29. Example work flow diagram.

electronic version of this is provided on the companion CD-ROM as *Stakeholder Involvement Matrix.doc.*

INFRASTRUCTURE

Organizational Policy Examples

The following information is provided for illustrative purposes. Each organization should focus on the development of policies and practices that best support its needs. Each organizational policy should include a top-level statement, any relevant responsibilities, and associated specific actions as desired. The following provides three examples of an organizational policy in support of configuration management activities.

Example 1

Individuals responsible for project leadership shall ensure that all software development, system integration, and system engineering projects conduct configuration management activities to maintain the integrity of their products and data. Each project's configuration management effort shall ensure the identification of configuration items (CIs) and associated baselines, the use of a change management process for managing changes to baselines, the support of process audits, and configuration status tracking and reporting.

Example 2

Our goal is to deliver the highest-quality products to our customers. To ensure that the integrity of a project's products is established and maintained throughout the

Table C-35. Stakeholder involvement matrix

Project Processes	Customer	Operations	Field Support	Program Management	System Engineering	Software Engineering	System Test	Integration and Test	Integrated Logistics Support	Quality Assurance	Configuration Management	Subcontracts	Help Desk	Suppliers
Proposal Generation	R			L	R	R	R	A	R			R		A
Project Planning	R		R	L	R	R	R	R	R			R		R
Project Monitoring and Control				L	R	R	A	R	R			R		A
Measures				R	L	R				O				
Risk Management	R			L	R	R	A	A	A					
Requirements Development	R	R	O		L	R	R		A	R				
Operations Telecons	R	R	O	O	L	R	O			O				A
Systems Engineering Review	R	R	R	R	L	R	R	R	R	R	R	R		R
Technical Exchange Meeting	R	A	R	R	L	R	R	A	A	O				A
Interface Control Working Group	R		R	R	L	R	R							A
Software Reviews					O	L	O		O	R				A
SW CCB				O	R	R	R	O	O	R	L			A
HW CCB				O	R		R	R		R	L			
Integration Testing					A	A	L			A				
Validation Testing		L	R		A	A	R			O				
Verification Testing			R				L			O				
Regression Testing			R				L			O				
Site Installation	R		R		A	A	A	L		R	A		A	
Leased Service Support				A	A	A	A	A	A		A	A	L	A
Second-Level Engineering Support					L	R	R		A		A		A	A
Subcontract Mgmt				A	A	A					A		L	A

Legend: R = Required, O = Optional, A = As Needed, L = Leader.

project's life cycle, this configuration management policy has three primary requirements:

1. Each project must have a CM Lead to develop and carry out the CM process. This individual is responsible for defining, implementing, and maintaining the CM plan.
2. Every project must have a CM plan identifying the products to be managed by the CM process, the tools to be used, the responsibilities of the CM team, and the procedures to be followed. The CM plan must be prepared in a timely manner.
3. Every project must have a configuration control board (CCB). The CCB authorizes the establishment of product baselines and the identification of configuration items/units. It also represents the interests of the Project Manager and all groups who may be affected by changes to the configuration. It reviews and authorizes changes to controlled baselines and authorizes the creation of products from the baselines.

Example 3

In order to ensure that delivered products contain only high-quality content, engineering projects shall:

- Establish and maintain baselines of all identified work products
- Track and control changes to all identified work products
- Establish and maintain the integrity of all baselines

Definition Form

A defined ISO 9001 activity should clearly state all inputs, entry criteria, associate activities, roles, measures, required verification, outputs, and exit criteria. Figure C-30 provides an example of a form that may be used in support of process definition and development.

Asset Library Catalog

IEEE Std 610.12 (R2002), IEEE Standard Glossary of Software Engineering Terminology, defines process as "A set of activities. A sequence of steps performed for a given purpose; for example, the software development process." [2] Similarly, ISO 9001 defines process as set of interrelated or interacting activities that transforms inputs into outputs. These inputs and outputs can be considered to be organizational assets.

Process assets are defined as ". . . artifacts that relate to describing, implementing, and improving processes" (e.g., policies, measurements, process descriptions, and process implementation support tools). The term "process assets" is used to in-

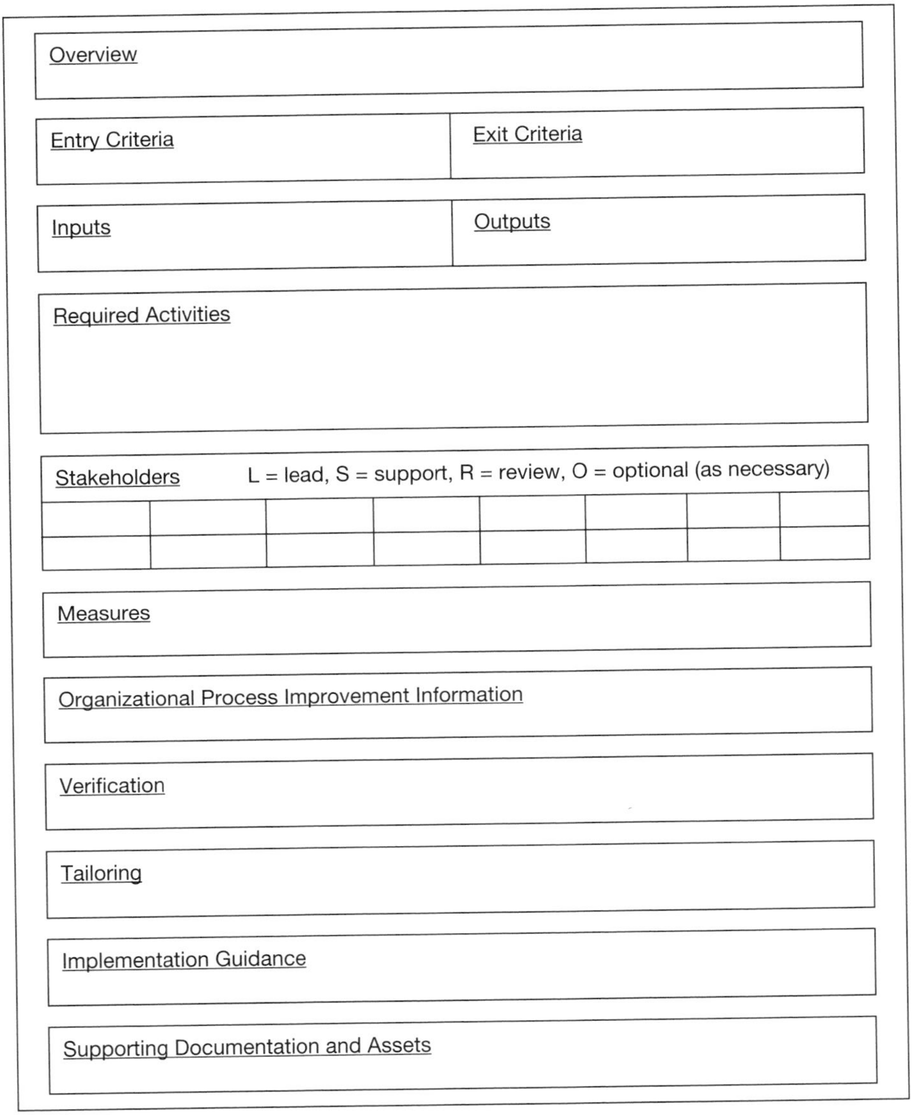

Overview	
Entry Criteria	Exit Criteria
Inputs	Outputs
Required Activities	
Stakeholders L = lead, S = support, R = review, O = optional (as necessary)	
Measures	
Organizational Process Improvement Information	
Verification	
Tailoring	
Implementation Guidance	
Supporting Documentation and Assets	

Figure C-30. Process definition.

dicate that these artifacts are developed or acquired to meet the business objectives of the organization, and they represent investments by the organization that are expected to provide current and future business value." [96]

Organizational assets include the following [96]:

- The organization's set of standard processes, including the process architectures and process elements

- Descriptions of life cycle models approved for use
- Guidelines and criteria for tailoring the organization's set of standard processes
- The organization's measurement repository
- The organization's asset library

An organization's asset library may, therefore, be described as a library of information used to store process assets that may be used in the support of the definition, implementation, and management of processes within an organization. Some examples of this type of documentation include policies, processes, procedures, checklists, lessons learned, templates and plans, referenced standards, and training materials.

IMPROVEMENT

Organizational Improvement Checklist

The IDEAL (initiating, diagnosing, establishing, acting, and learning) model is an organizational improvement model that was originally developed to support CMM®-based software process improvement [87]. It serves as a road map for initiating, planning, and implementing improvement actions. This model can serve to lay the groundwork for a successful improvement effort (initiate), determine where an organization is in reference to where it may want to be (diagnose), plan the specifics of how to reach goals (establish), define a work plan (act), and apply the lessons learned from past experience to improve future efforts (learn).

This model serves as the basis for CMMI(r) process improvement. The information presented in Table C-36 provides a matrix of IDEAL^{SM} phases, activities, details, and deliverables in support of process improvement and may be used as a reference checklist in support of process improvement activities.

Organization Process Appraisal Checklist

The information provided in Table C-37 may be used in support of the organization process appraisals. An electronic version of this checklist is provided on the companion CD-ROM, identified as *Process Appraisal Checklist.doc*.

Lessons Learned

The documentation of lessons learned facilitates individual project knowledge sharing so as to benefit the entire organization. A successful lessons-learned program can support the repetition of desirable outcomes and avoid undesirable outcomes. Lessons learned should not only be used as a closeout activity; the continual recording of lessons learned throughout the project is the best way to ensure that informa-

Table C-36. Organizational process improvement checklist using IDEAL^{SM} [98]

IDEAL^{SM} phases (5)	IDEAL^{SM} activities (14)	IDEAL^{SM} details	Organization milestone details/deliverables
	Stimulus for change	Know what prompted this particular change: • Reaction to unanticipated events/ circumstances • Edict from "on high" • Enactment of change as part of proactive continuous improvement approach	Business results Problem reports
Initiating	Set context	Identify where this change fits within the larger organizational context: • Within the organization's business strategy	Business goals and objectives
		Specify business goals and objectives that will be realized or supported: • The organization's core mission • Business goals and objectives • A coherent vision for the future • A strategy for achieving the vision • Reference models identify impact on other initiatives and on-going work	
	Build sponsorship	Secure the support required to give the change a reasonable chance of succeeding. Effective sponsors will: • Give personal attention to the effort • Stick with the change through difficult times • Commit scarce resources to the effort • Change their own behavior to be consistent with the change being implemented • Modify the reward system.	Sponsor(s) signed up
	Charter infrastructure	Adjust relevant organizational systems to support the effort. Establish an infrastructure for managing specifics of implementation: • An oversight group (senior-level people) • A change agency group (staff to the oversight body) • One or more technical working groups (TWGs)	Engineering process group (EPG) charter

Table C-36. *Continued*

IDEAL^{SM} phases (5)	IDEAL^{SM} activities (14)	IDEAL^{SM} details	Organization milestone details/deliverables
Diagnosing	Characterize current and desired states	Identify the organization's current state with respect to a related standard or reference model: • A reference model (e.g., capability maturity model) • Standardized appraisal instruments and methodologies associated with the model • Business objectives aligned with the change effort Identify the organization's desired state against the same standard or model: • Analyze the gap between the current and desired states • Determine aspects of current state to retain • Determine the work needed to reach the desired state • Align recommendations with business objectives • Identify potential projects that you may pilot or test	Process Appraisal Plan, Assessment Plan
	Develop recommendations	Recommend actions that will move the organization from where it is to where you want it to be. Identify potential barriers to the effort	Recommendation report
Establishing	Set priorities	Develop priorities among the recommended actions on the basis of such things as: • Availability of key resources • Dependencies between the actions recommended • Funding Organizational realities: • Resources needed for change are limited • Dependencies exist between recommended activities • External factors may intervene • Organization's strategic priorities must be honored	

(*continued*)

Table C-36. *Continued*

IDEAL^{SM} phases (5)	IDEAL^{SM} activities (14)	IDEAL^{SM} details	Organization milestone details/deliverables
Establishing (*cont.*)	Develop approach	Develop an approach that reflects priorities and current realities: • Strategy to achieve vision (set during initiating phase) • Specifics of installing the new technology • New skills and knowledge required of the people who will be using the technology • Organizational culture • Potential pilot projects • Sponsorship levels • Market forces	
	Plan actions	Develop plans to implement the chosen approach: • Deliverables, activities, and resources • Decision points • Risks and mitigation strategies • Milestones and schedule • Measures, tracking, and oversight	Organizational process improvement plan; process action plan (PAP)
Acting	Create solution	Bring together everything available to create a "best-guess" solution specific to organizational needs: • Existing tools, processes, knowledge, skills, etc. • New knowledge, information, etc. • Outside help Solution should: • Identify performance objectives • Finalize the pilot/test group • Construct the solution material and training with pilot/test group • Develop draft plans for pilot/test and implementation Technical working groups develop the solution with pilot/test group.	Revised process description and/or process asset
	Pilot/test solution	Put the solution in place on a trial basis to learn what works and what does not: • Train the pilot or test group • Perform the pilot or test • Gather feedback	Process lessons learned

Table C-36. *Continued*

IDEAL^{SM} phases (5)	IDEAL^{SM} activities (14)	IDEAL^{SM} details	Organization milestone details/deliverables
Acting (*cont.*)	Refine solution	Revisit and modify the solution to incorporate new knowledge and understanding. Iterations of pilot test and refine may be necessary before arriving at a solution that is deemed satisfactory.	Revised process description and/or process asset
	Implement solution	When solution is deemed workable and sufficient, implement it throughout the organization: • Adjust the plan as necessary • Execute the plan • Conduct postimplementation analysis • Various approaches may be used: top-down, lateral, or staged	Process lessons learned
Learning.	Analyze and validate	Determine what was actually accomplished by the effort: • In what ways did it/did it not accomplish its intended purpose? • What worked well? • What could be done more effectively or efficiently? Compare what was accomplished with the intended purpose for undertaking the change. Summarize/roll up lessons learned regarding processes used to implement IDEAL.	Change in measurements
	Propose future actions	Develop recommendations concerning management of future change efforts using IDEAL: • Improving an organization's ability to use IDEAL • Address a different aspect of the organization's business	

Table C-37. Organization process appraisal checklist

Checklist	Dates + resources
1. Identify Appraisal Scope	
a. Appraisal input subject to sponsor approval	
2. Develop Appraisal Plan (using this checklist)	
a. Select appraisal class	
b. Identify appraisal scope	
c. Select appraisal team members	
d. Develop and publish appraisal plan	
3. Prepare Organization for Appraisal	
a. Prepare and train appraisal team	
b. Brief appraisal participants	
c. Data collection and analysis	
d. Administer questionnaire	
e. Conduct initial document review	
f. Remediate, consolidate, triage responses	
g. Conduct opening meeting	
h. Interview project leaders	
i. Consolidate information	
j. Interview middle managers	
k. Consolidate information	
l. Interview FAR groups	
m. Consolidate information	
n. Verify and validate objective evidence	
o. Prepare draft findings	
p. Present draft findings	
q. Consolidate information	
4. Rate and prep final findings	
5. Present final findings	
6. Conduct executive session	
7. Present final findings to appraisal participants	
8. Wrap-up appraisal	

tion is accurately recorded. Figure C-31 is provided as a lessons-learned template. Please refer to the companion CD-ROM for an electronic version of this template, entitled *Lessons Learned.doc.*

Measures Definition for Organizational Processes

IEEE Std. 982.1, IEEE Standard Dictionary of Measures to Produce Reliable Software; IEEE 1061, IEEE Standard for a Software Quality Metrics Methodology; IEEE Std. 14143.1, Implementation Note for IEEE Adoption of ISO/IEC 14143, Information Technology—Software Measurement—Functional Size Measurement—Part 1: Definition of Concepts; ISO/IEC 15939, Information Technology—Software Engineering—Software Measurement Process; and ISO 9001 materials were

---

[Project Name] Lessons Learned
Revision [xx], date

Team Meeting
   During each project team meeting, discuss what strategies contributed to success as well as areas of potential improvement. Enter your conclusions below (insert rows as needed):

Strategies and Processes that led to Success

Date   Description

Areas of Potential Improvement

Date   Description

Project Close Out Discussion
   All stakeholders should participate in the lessons-learned closeout discussion. Use the questions below to trigger discussion and summarize findings.

1. List the project's three biggest successes

Description   Contributing Factors

2. Other relevant successes

Description   Contributing Factors

3. List improvement items and proposed strategies

Description   Improvement Strategy

4. Additional Comments

Document Approval
   The following stakeholders have reviewed the information contained in this Project lessons learned document and agree with its content.

Name   Title   Signature   Date

Figure C-31. Example lessons-learned template. Adapted from a template written by Covansys for the City of Raleigh, NC, Enterprise PMO. Generic format by CVR-IT (www.cvr-it.com). May be used freely if referenced.

used as to develop the following suggestions for measures in support of organizational processes.

Measures are often classified as either process or product measures. Process measures are based upon characteristics associated with the development process or environment. Product measures are measures associated with the software product being developed. Examples of these types of measures are provided in Table C-38.

The basic information in support of process measurement that should be collected is [80]:

- Released defect levels
- Product development cycle time
- Schedule and effort estimating accuracy
- Reuse effectiveness
- Planned and actual cost

TRAINING

Figure C-32 is a form that may be used to request training. Project managers should note training needs within their project. Use as many forms as required to document all training needs. Training managers may use this form to analyze, evaluate, validate, and prioritize training requests. An electronic version, *Training Request Form.doc,* is provided on the companion CD-ROM.

Training Log

Table C-39 is a form that may be used to document training. Project managers or individuals delegated with this responsibility should record all completed training. Training managers may use this form to analyze, evaluate, validate, and prioritize training requests. An electronic version, *Training Log.doc,* is included on the companion CD-ROM.

Table C-38. Process versus product measures

Process Measures
  Estimated and actual duration (calendar time): track for individual tasks, project milestones, and process support functions.
  Estimated and actual effort (labor hours): track for individual tasks, project milestones, and process support functions.
Product Measures
  Count lines of code, function points, object classes, and numbers of requirements.
  Time spent in development and maintenance activities.
  Count the number of defects identifying their type, severity, and status.

Project Name: _____ Project Manager: _____

Project Location: _____ Date: _____

Will project contract pay for requested training?   Yes   No   Labor only

Describe the training requirement:

Purpose and date training required:

Number of employees to be trained:

Can employees be released for training during work hours?   Yes   No

Are subject matter experts available?   Yes   No

If so, names:

Project training requirement priority (5 is the highest, 1 is the lowest)

5   4   3   2   1

Recommended provider (if known)

Recommended course title (if known)

Recommended action:

Program Manager: _____ Date: _____

Division Manager: _____ Date: _____

Status: (to be completed by training manager)

Training manager: _____ Date: _____

SBU training requirement priority (5 is the highest, 1 is the lowest)

5   4   3   2   1

Action taken:

Invalid requirement*   _____Valid requirement_____

Placed on training schedule_____ No action taken*_____

Funding identified: _____

Training course identified: _____

Placed on development schedule: _____

Provider: _____

Course title: _____ Date: _____

*Remarks:

Figure C-32. Training request form.

Table C-39. Training log

Trainee Name	Date	Location	Description of Training	No. Hours	ID #	Comments

Appendix D

ISO/IEC Guidance

ISO 9001:2000 MAPPING TO ISO/IEC STANDARDS

The following reference table is from ISO/IEC 90003 [48]—Additional Guidance in the Implementation of ISO 9001:2000, and is also available in the ISO/IEC JTC 1/SC 7 standards as well as ISO 9000, ISO 9001, and ISO 9004. This table reflects a mapping from ISO 9001:2000 to related ISO/IEC standards.

ISO 9001:2000	ISO/IEC
4. Quality management system	
4.1. General requirements	ISO/IEC 12207:1995 and Amd.1
	ISO/IEC TR 15271 Annex C
4.2. Documentation requirements	ISO/IEC 12207:1995 and Amd.1 6.1, F.2.1
5. Management responsibility	
5.1. Management commitment	
5.2. Customer focus	
5.3. Quality policy	
5.4. Planning	
5.4.1. Quality objectives	ISO/IEC TR 15504 Pt 1
5.4.2. Quality management system planning	
5.5. Responsibility, authority, and communication	
5.6. Management review	
6. Resource management	
6.1. Provision of resources	
6.2. Human resources	
6.2.1. General	ISO/IEC 12207:1995 and Amd.1 F.3.3.1, F.3.32
6.2.2. Competence, awareness, and training	*(continued)*

ISO 9001:2000	ISO/IEC
6.3. Infrastructure	ISO/IEC 12207:1995 and Amd.1 7.2, F.3.2 ISO/IEC 14598 Pt 2, Pt 3 ISO/IEC 14102
6.4. Work environment	
7. Product realization	
7.1. Planning of product realization	ISO/IEC 12207:1995 and Amd.1 5.2.4, 5.3.1, 6.1 to 6.8, F.2 ISO/IEC 9126 Pt 1 ISO/IEC 14598 Pt 2 ISO/IEC TR 15846 6.2 ISO/IEC TR 16326 6.2.2
7.2. Customer-related processes	
7.2.1. Determination of requirements related to the product	ISO/IEC 12207:1995 and Amd.1 5.3.2 to 5.3.4, F.1.3.1, 2, 4 ISO/IEC 9126 Pt 1 ISO/IEC 12119 ISO/IEC 15026
7.2.2. Review of requirements related to the product	ISO/IEC 12207:1995 and Amd.1 5.2.1, 5.2.6, 6.4.2.1, 6.6, F.3.1.5
7.2.3. Customer communication	ISO/IEC 12207:1995 and Amd.15.2.5.6, 5.2.6, 5.2.7, 6.6, F.1.4.2 ISO/IEC 14764 6.8.1, 7.3.3, 8.2, 8.2.3
7.3. Design and development	ISO/IEC 12207:1995 and Amd.1 F.1.3.4, F.1.3.5 ISO/IEC 12119 ISO/IEC 6592 ISO/IEC 14143-1 ISO/IEC 15910
7.3.1. Design and development planning	ISO/IEC 12207:1995 and Amd.1 5.2.4, 5.3.1 ISO/IEC TR 16326 6.2.2
7.3.2. Design and development inputs	ISO/IEC 9126 Pt 1
7.3.3. Design and development outputs	ISO/IEC 12207:1995 and Amd.1 5.3.5 to 5.3.7
7.3.4. Design and development review	ISO/IEC 12207:1995 and Amd.1 5.3.4.2, 5.3.5.6, 5.3.6.7, 6.6.3, F.2.6 ISO/IEC TR 15271 Annex A
7.3.5. Design and development verification	ISO/IEC 12207:1995 and Amd.1 5.3, 6.4, F.1.3, F.2.4
7.3.6. Design and development validation	ISO/IEC 12207:1995 and Amd.1 5.3, 6.5, F.1.3, F.2.5 ISO/IEC 14598 Pt 3, Pt 5

ISO 9001:2000	ISO/IEC
7.3.7. Control of design and development changes	ISO/IEC 12207:1995 and Amd.1 5.5.2 5.5.3, 6.1, 6.2, F.2.1, F.2.2
7.4. Purchasing	ISO/IEC 9126 Pt 1
	ISO/IEC 14598 Pt 4
	ISO/IEC 14143-1
7.4.1. Purchasing process	ISO/IEC 12207:1995 and Amd.1 5.1, F.1.1
	ISO/IEC TR 15504 Pt 8
7.4.2. Purchasing information	ISO/IEC 12207:1995 and Amd.1 5.1.2, F.1.1.1
7.4.3. Verification of purchased product	ISO/IEC 12207:1995 and Amd.1 5.1.5, F.1.1.4
7.5. Production and service provision	ISO/IEC 9126 Pt 1
	ISO/IEC TR 15846
	ISO/IEC 14764
	ISO/IEC 15910
7.5.1. Control of production and service provision	ISO/IEC 12207:1995 and Amd.1 5.3.12, 5.4.4, 5.5, 6.3.3, 6.8, F.1.3.11, F.1.4.2, F.1.5, F.2.8
7.5.2. Validation of processes for production and service provision	
7.5.3. Identification and traceability	ISO/IEC 12207:1995 and Amd.1 6, F.2.2
	ISO/IEC TR 15846 7 to 12
	ISO 10007
7.5.4. Customer property	
7.5.5. Preservation of product	
7.6. Control of monitoring and measuring devices	
8. Measurement, analysis, and improvement	
8.1. General	ISO/IEC 12207:1995 and Amd.1 7, F.3.3
	ISO/IEC 9126 Pt 2, Pt 3
	ISO/IEC 14598 Pt 2
	ISO/IEC TR 15504 Pt 1
	ISO/IEC 15939 5
8.2. Monitoring and measurement	
8.2.1. Customer satisfaction	ISO/IEC 9126 Pt 4
8.2.2. Internal audit	ISO/IEC 12207:1995 and Amd.1 6.3, 6.7, F.2.3, F.2.7
8.2.3. Monitoring and measurement of processes	ISO/IEC 12207:1995 and Amd.1 7.3.2, 7.3.3, F.3.3.2
	ISO/IEC TR 15504 Pt 1, Pt 3
	ISO/IEC 15939 5 *(continued)*

ISO 9001:2000	ISO/IEC
8.2.4. Monitoring and measurement of product	ISO/IEC 12207:1995 and Amd.1 5.3, F.1.3 ISO/IEC 9126 Pt 1 ISO/IEC 14598 Pt 3, Pt 5
8.3. Control of nonconforming product	ISO/IEC 12207:1995 and Amd.1 6.2, 6.8, F.2.2, F.2.8 ISO/IEC 12119 ISO/IEC TR 15846
8.4. Analysis of data	ISO/IEC 15939 5.4 ISO/IEC 14143-1
8.5. Improvement	
8.5.1. Continual improvement	ISO/IEC 12207:1995 and Amd.1 7.3, F.3.3 ISO/IEC TR 15504 Pt 7
8.5.2. Corrective action	ISO/IEC 12207:1995 and Amd.1 6.8, F.2.8
8.5.3. Preventive action	ISO/IEC 12207:1995 and Amd.1 7.3.2, F.3.3.2 ISO/IEC TR 15504 Pt 3

Appendix **E**

ISO/IEC 90003 Mapping to ISO/IEC 12207

The following reference table is from ISO/IEC 90003 [48]—Additional Guidance in the Implementation of ISO/IEC 12207, and is available from ISO/IEC 12207. This table reflects a mapping from ISO/IEC 90003 to related ISO/IEC 12207.

ISO/IEC 90003:2003 Reference	ISO/IEC 12207:1995 Reference
7.1.2. Planning of product realization	
a) Inclusion of, or reference to, the plans for development (see 7.3.1)	5.3.1.4. The developer shall develop plans for conducting the activities of the development process.
b) Quality requirements related to the product and/or processes	5.2.4.5 (d). Management of the quality characteristics of the software products or services. Separate plans for quality may be developed.
	5.2.4.5 (e). Management of the safety, security, and other critical requirements of the software products or services. Separate plans for safety and security may be developed
c) Quality management system tailoring and/or identification of specific procedures and instructions, appropriate to the scope of the quality manual and any stated exclusions (see ISO 9001:2000, 1.2)	Note: Not addressed specifically by ISO/IEC 12207 since this relates to the organization level.
d) Project-specific procedures and instructions, such as software test specifications detailing plans, designs, test cases, and procedures for unit, integration, system, and acceptance testing (see 8.2.4)	Note: Covered in 5.3.1.3 of the development process and 6.1, documentation process, which is invoked across development activities.

(*continued*)

ISO/IEC 90003:2003 Reference	ISO/IEC 12207:1995 Reference
7.1.2. Planning of product realization (*cont.*)	
e) Methods, life cycle model(s), tools, programming language conventions, libraries, frameworks, and other reusable assets to be used in the project	6.3.1.3 (a). Quality standards, methodologies, procedures, and tools for performing the quality assurance activities (or their references in organization's official documentation).
	5.2.4.5 (b). Engineering environment (for development, operation, or maintenance, as applicable), including test environment, library, equipment, facilities, standards, procedures, and tools.
f) Criteria for starting and ending each project stage	Note: Covered in 6.6, joint review process.
g) Types of review, and other verification and validation activities to be carried out (see 7.3.4, 7.3.5 and 7.3.6)	5.2.4.5 (g). Quality assurance (see 6.3).
	5.2.4.5 (h). Verification (see 6.4) and validation (see 6.5); including the approach for interfacing with the verification and validation agent, if specified.
	6.3.1.3 (e). Selected activities and tasks from supporting processes, such as Verification (6.4), Validation (6.5), Joint Review (6.6), Audit (6.7), and Problem Resolution (6.8).
h) Configuration management procedures to be carried out (see 7.5.3)	Note: Covered in 6.2, configuration management process, and invoked in 5.3.1.2 (b) in development.
i) Monitoring and measurement activities to be carried out	Note: Monitoring is part of 5.2.5.3 in the supply process and measurement is part of product and process assurance in 6.3.3.5
j) The person(s) responsible for approving the outputs of processes for subsequent use	Note: Where outputs of processes are documents this is covered by 6.1.2.3 in the documentation process.
k) Training needs in the use of tools and techniques, and scheduling of the training before the skill is needed	5.2.4.5 (o). Training of personnel (see 7.4).
l) Records to be maintained (see 4.2.4)	6.3.1.3 (c). Procedures for identification, collection, filing, maintenance, and disposition of quality records.

ISO/IEC 90003:2003 Reference	ISO/IEC 12207:1995 Reference
7.1.2. Planning of product realization (*cont.*)	
m) Change management, such as for resources, timescale and contract changes	Note: Covered in 5.1.3.5 as a change control mechanism between the acquirer and supplier.
7.3.1. Design and development planning	5.2.4.5 (o). Training of personnel (see 7.4).
a) The activities of requirements analysis, design and development, coding, integration, testing, installation, and support for acceptance of software products; this includes the identification of, or reference to:	6.3.1.3 (e). Selected activities and tasks from supporting processes, such as Verification (6.4), Validation (6.5), Joint Review (6.6), Audit (6.7), and Problem Resolution (6.8).
1) Activities to be carried out	
2) Required inputs to each activity	Note: Specific requirements are included in activities and tasks of 5.3 development, 5.4 operation and 5.5 maintenance processes, rather than as a planning activity.
3) Required outputs from each activity	
4) Verification required for each activity output [as 7.1.2 (g)—see also 7.3.5]	
5) Management and supporting activities to be carried out	
6) Required team training [as 7.1.2 (k)]	
b) Planning for the control of product and service provision	Note: This term from ISO 9001:2000 is not used in ISO/IEC 12207. It equates to all release, delivery, and postdelivery activities, including software installation (5.3.12), software acceptance support (5.3.13), and including operation (5.4), and maintenance (5.5) processes, as well as configuration management (6.2). These are specific requirements of ISO/IEC 12207 subclauses, rather than a planning activity.
c) The organization of the project resources, including the team structure, responsibilities, use of suppliers and material resources to be used	5.2.4.5 (a). Project organizational structure and authority and responsibility of each organizational unit, including external organizations.

(*continued*)

ISO/IEC 90003:2003 Reference	ISO/IEC 12207:1995 Reference

7.3.1. Design and development planning (*cont.*)

  d) Organizational and technical interfaces between different individuals or groups, such as subproject teams, suppliers, partners, users, customer representatives, and quality assurance representative (see 7.3.1.4)

  5.2.4.5 (i). Acquirer involvement; that is, by such means as joint reviews (see 6.6), audits (see 6.7), informal meetings, reporting, modification and change; implementation, approval, acceptance, and access to facilities.

  5.2.4.5(j). User involvement; by such means as requirements setting exercises, prototype demonstrations, and evaluations.

  e) The analysis of the possible risks, assumptions, dependencies and problems associated with the design and development

  5.2.4.5 (k). Risk management; that is, management of the areas of the project that involve potential technical, cost, and schedule risks.

Note: Problems are covered by the problem resolution process (6.8).

  f) The schedule identifying:
    1) The stages of the project [see also 7.1.2 (j)]
    2) The work breakdown structure
    3) The associated resources and timing
    4) The associated dependencies
    5) The milestones
    6) Verification and validation activities [as 7.1.2 (g)]

  5.2.4.5 (c). Work breakdown structure of the life cycle processes and activities, including the software products, software services, and nondeliverable items, to be performed together with budgets, staffing, physical resources, software size, and schedules associated with the tasks.

  5.2.4.5 (n). Means for scheduling, tracking, and reporting.

  6.3.1.3 (d). Resources, schedule, and responsibilities for conducting of the quality assurance activities.

  g) The identification of:
    1) Standards, rules, practices and conventions, methodology, life cycle model and statutory and regulatory requirements [as 7.1.2 (d) and (e)]
    2) Tools and techniques for development, including the qualification of, and configuration controls placed on, such tools and techniques

  5.3.1.4. The developer shall develop plans for conducting the activities of the development process. The plans should include specific standards, methods, tools, actions, and responsibility associated with the development and qualification of all requirements including safety and security. If necessary, separate plans may be developed. These plans shall be documented and executed.

ISO/IEC 90003:2003 Reference	ISO/IEC 12207:1995 Reference

7.3.1. Design and development planning (*cont.*)

  g) The identification of: (*cont.*)

    3) Facilities, hardware, and software for development

    4) Configuration management practices [as 7.1.2 (h)]

    5) Method of controlling non-conforming software products

    6) Methods of control for software used to support development

    7) Procedures for archiving, backup, recovery, and controlling access to software products

    8) Methods of control for virus protection

    9) Security controls

5.2.4.5 (m). Approval required by such means as regulations, required certifications, proprietary, usage, ownership, warranty and licensing rights.

  h) The identification of related planning (including planning of the system) addressing topics such as quality (see 7.1), risk management, configuration management, supplier management, integration, testing (see 7.3.6), release management, installation, training, migration, maintenance, re-use, communication, and measurement

5.2.4.5 (g). Quality assurance (see 6.3).

5.2.4.5 (k). Risk management; that is, management of the areas of the project that involve potential technical, cost, and schedule risks.

5.2.4.5 (l). Security policy; that is, the rules for need to know and access to information at each project organization level.

Note: ISO 9001:2000 does not require contract review procedures but includes in 7.2.2 (a) the requirement to review requirements before acceptance of a contract.

6.3.1.3 (b). Procedures for contract review and coordination thereof.

7.3.4 Design and development review
Review of design and development should be performed in accordance with planned arrangements. The elements of the review to be considered are the following:

6.6.1.1. Periodic reviews shall be held at predetermined milestones as specified in the project plan(s).

(continued)

ISO/IEC 90003:2003 Reference	ISO/IEC 12207:1995 Reference

7.3.4 Design and development review
(*cont.*)

a) What is to be reviewed, when and the type of review, such as demonstrations, formal proof of correctness, inspections, walk-throughs, and joint reviews

6.6.1.3. The parties should agree on the following items at each review: meeting agenda, software products (results of an activity), and problems to be reviewed; scope and procedures; and entry and exit criteria for the review.

b) What functional groups would be concerned in each type of review and, if there is to be a review meeting, how it is to be organized and conducted?

6.6.1.2. All resources required to conduct the reviews shall be agreed on by the parties. These resources include personnel, location, facilities, hardware, software, and tools. Also 6.6.1.3 as above.

c) What records have to be produced, e.g., meeting minutes, issues, problems, actions, and action status?

6.6.1.4. Problems detected during the reviews shall be recorded and entered into the Problem Resolution Process (6.8) as required.

6.6.1.5. The review results shall be documented and distributed. The reviewing party will acknowledge to the reviewed party the adequacy (for example, approval, disapproval, or contingent approval) of the review results.

d) The methods for monitoring the application of rules, practices and conventions to ensure requirements are met

6.6.3.1. Technical reviews shall be held to evaluate the software products or services under consideration and provide evidence that they comply with their standards and specifications.

e) What has to be done prior to the conduct of a review, such as establishment of objectives, meeting agenda, documents required, and roles of review personnel

6.6.1.3. The parties should agree on the following items at each review: meeting agenda, software products (results of an activity) and problems to be reviewed; scope and procedures; and entry and exit criteria for the review.

6.6.1.2. All resources required to conduct the reviews shall be agreed on by the parties. These resources include personnel, location, facilities, hardware, software, and tools.

ISO/IEC 90003:2003 Reference	ISO/IEC 12207:1995 Reference

7.3.4 Design and development review (*cont.*)

f) What has to be done during the review, including the techniques to be used and guidelines for all participants

6.6.1.3. The parties should agree on the following items at each review: . . . scope and procedures. . . .

g) The success criteria for the review

6.6.1.3. (end) . . . and entry and exit criteria for the review.

h) What follow-up activities are used to ensure that issues identified at the review are resolved?.

6.6.1.6. The parties shall agree on the outcome of the review and any action item responsibilities and closure criteria.

Note: ISO 9001:2000 does not distinguish between project management and technical reviews. This is often very useful for software projects. Although not included in detail in the guidance for ISO 9001:2000 in this International Standard, it may be appropriate and useful to use the ISO/IEC 12207 view of these separate review mechanisms.

6.6.2. Project management reviews

6.6.2.1. Project status shall be evaluated relative to the applicable project plans, schedules, standards, and guidelines. The outcome of the review should be discussed between the two parties and should provide for the following:

a) Making activities progress according to plan, based on an evaluation of the activity or software product status

b) Maintaining global control of the project through adequate allocation of resources

c) Changing project direction or determining the need for alternate planning

d) Evaluating and managing the risk issues that may jeopardize the success of the project

Note: ISO 9001:2000 does not distinguish between project management and technical reviews. This is often very useful for software projects. Although not included in detail in the guidance for ISO 9001:2000 in this International Standard, it may be appropriate and useful to use the ISO/IEC 12207 view of these separate review mechanisms.

6.6.3. Technical reviews

6.6.3.1. Technical reviews shall be held to evaluate the software products or services under consideration and provide evidence that:

a) They are complete

b) They comply with their standards and specifications

(continued)

ISO/IEC 90003:2003 Reference	ISO/IEC 12207:1995 Reference
7.3.4 Design and development review (*cont.*)	c) Changes to them are properly implemented and affect only those areas identified by the Configuration Management Process (6.2) d) They are adhering to applicable schedules e) They are ready for the next activity f) The development, operation, or maintenance is being conducted according to the plans, schedules, standards, and guidelines of the project

Appendix F

CD ROM Reference Summary

The following table is provided as a summary reference of the template items found on the companion CD-ROM.

Process	CD-ROM Template
Acquisition	Acquisition Strategy Checklist
Acquisition	Software Acquisition Plan
Acquisition	Supplier Checklist
Acquisition	Supplier Performance Standards
Acquisition	Decision Matrix
Audit	Software Measurement and Metrics Plan
Configuration Management	Software Configuration Plan
Configuration Management	CCB Charter
Configuration Management	CCB Letter of Authorization
Development	Small Software Project Management Plan
Development	Interface Control Document
Development	Software Transition Plan
Development	Software Users Manual
Development	System Integration Test Report
Development	Unit Test Report
Development	Software Design Document
Development	Software Requirements Specification
Development	System Requirements Specification
Development, Management	Software Project Management Plan
Development, Management	Small Software Project Management Plan
Development, Management	Software Requirements Management Plan
Development, Validation	System Integration Test Plan
Improvement	Lessons Learned
Joint Review	Open Issues List
Maintenance and Configuration Management	Baseline Change Request Risk Management Plan

(*continued*)

Process	CD-ROM Template
Management	Stakeholder Involvement Matrix
Quality Assurance	Process Appraisal Checklist
Quality Assurance	Software Quality Assurance Plan
Quality Assurance	SQA Inspection Log
Supplier	ConOps Document
Training	Training Log
Training	Training Plan
Training	Training Request Form
Validation	Software Test Plan
Validation	System Test Plan
Verification	Inspection Log Defect Summary
Verification	Inspection Report

References

IEEE PUBLICATIONS

[1] IEEE/ANSI. IEEE Guide to Software Configuration Management. ANSI/IEEE Std 1042-1987, IEEE Press, New York, NY, 1987.

[2] IEEE Standard Glossary of Software Engineering Terminology, IEEE Std 610.12-1990 (Sep 28), Reaffirmed Sep 2002, IEEE Press, New York, NY, 2002.

[3] IEEE Standard for Software Quality Assurance Plans, IEEE Std 730-2002 (Sep), IEEE Press, New York, NY, 2002.

[4] IEEE Standard for Software Configuration Management Plans, IEEE Std 828-1998 (Jun 25), IEEE Press, New York, NY, 1998.

[5] IEEE Standard for Software Test Documentation, IEEE Std 829-1998 (Sep 16), IEEE Press, New York, NY, 1998.

[6] IEEE Recommended Practice for Software Requirements Specifications, IEEE Std 830-1998 (Jun 25), IEEE Press, New York, NY, 1998.

[7] IEEE Standard Dictionary of Measures to Produce Reliable Software, IEEE Std 982.1-1988 (Jun 9), IEEE Press, New York, NY, 1988.

[8] An American National Standard—IEEE Standard for Software Unit Testing, ANSI/IEEE Std 1008-1987(R1993), Reaffirmed Dec. 2002, IEEE Press, New York, NY, 2002.

[9] IEEE Standard for Software Verification and Validation, IEEE Std 1012-1998 (Mar 9), IEEE Press, New York, NY, 1998.

[10] Supplement to IEEE Standard for Software Verification and Validation: Content Map to IEEE/EIA 12207.1-1996, IEEE Std 1012a-1998 (Sep 16), IEEE Press, New York, NY, 1998.

[11] IEEE Recommended Practice for Software Design Descriptions, IEEE Std 1016-1998 (Sep 23), IEEE Press, New York, NY, 1998.

[12] IEEE Standard for Software Reviews, IEEE Std 1028-1997 (Mar 4), Reaffirmed Sep 2002, IEEE Press, New York, NY, 2002.

[13] IEEE Standard Classification for Software Anomalies, IEEE Std 1044-1993 (Dec 2), Reaffirmed Sep 2002, IEEE Press, New York, NY, 2002.

[14] IEEE Standard for Software Productivity Metrics, IEEE Std 1045-1992 (Sep 17), Reaffirmed Dec 2002, IEEE Press, New York, NY, 2002.

[15] IEEE Standard for Software Project Management Plans, IEEE Std 1058-1998 (Dec 8), IEEE Press, New York, NY, 1998.

[16] IEEE Standard for a Software Quality Metrics Methodology, IEEE Std 1061-1998 (Dec 8), IEEE Press, New York, NY, 1998.

[17] IEEE Recommended Practice for Software Acquisition, IEEE Std 1062-1998 Edition (Dec 2), Reaffirmed Sep 2002, IEEE Press, New York, NY, 2002.

[18] IEEE Standard for Software User Documentation, IEEE Std 1063-2001 (Dec 5), IEEE Press, New York, NY, 2001.

[19] IEEE Standard for Developing Software Life Cycle Processes, IEEE Std 1074-1997 (Dec 9), IEEE Press, New York, NY, 1997.

[20] IEEE Standard Reference Model for Computing System Tool Interconnections, IEEE Std 1175-1991 (Dec 5), IEEE Press, New York, NY, 1991.

[21] IEEE Guide for CA Software Engineering Tool Interconnections—Classification and Description, IEEE Std 1175.1-2002 (Nov. 11), IEEE Press, New York, NY, 2002.

[22] IEEE Standard for Software Maintenance, IEEE Std 1219-1998 (Jun 25), IEEE Press, New York, NY, 1998.

[23] IEEE Standard for the Application and Management of the Systems Engineering Process, IEEE Std 1220-1998 (Dec 8), IEEE Press, New York, NY, 1998.

[24] IEEE Standard for Software Safety Plans, IEEE Std 1228-1994 (Mar 17), Reaffirmed Dec 2002, IEEE Press, New York, NY, 2002.

[25] IEEE Guide for Developing System Requirements Specifications, IEEE Std 1233, 1998 Edition (Apr 17), Reaffirmed Sep 2002, IEEE Press, New York, NY, 2002.

[26] IEEE Standard for Functional Modeling Language—Syntax and Semantics for IDEF0, IEEE Std 1320.1-1998 (Jun 25), IEEE Press, New York, NY, 1998.

[27] IEEE Standard for Conceptual Modeling Language Syntax and Semantics for IDEF1X 97 (IDEF object), IEEE Std 1320.2-1998 (Jun 25), IEEE Press, New York, NY, 1998. IEEE Guide for Information Technology-System Definition-Concept of Operations (ConOps) Document, IEEE Std 1362-1998 (Mar 19), IEEE Press, New York, NY, 1998.

[28] IEEE Standard for Information Technology—System Definition—Concept of Operations (ConOps) Document, IEEE Std 1362-1998 (Mar 19), IEEE Press, New York, NY, 1998. IEEE Standard for Information Technology—Software Reuse—Data Model for Reuse Library Interoperability: Basic Interoperability Data Model (BIDM), IEEE Std 1420.1-1995 (Dec 12), Reaffirmed Jun 2002, IEEE Press, New York, NY, 2002.

[29] Supplement to IEEE Standard for Information Technology—Software Reuse—Data Model for Reuse Library Interoperability: Asset Certification Framework, IEEE Std 1420.1a-1996 (Dec 10), Reaffirmed Jun 2002, IEEE Press, New York, NY, 2002.

[30] IEEE Trial-Use Supplement to IEEE Standard for Information Technology—Software Reuse—Data Model for Reuse Library Interoperability: Intellectual Property Rights Framework, IEEE Std 1420.1b-1999 (Jun 26), Reaffirmed Jun 2002, IEEE Press, New York, NY, 2002.

[31] IEEE Standard—Adoption of International Standard ISO/IEC 14102: 1995—Information Technology—Guideline for the Evaluation and Selection of CASoftware Engineering tools, IEEE Std 1462-1998 (Mar 19), IEEE Press, New York, NY, 1998.

[32] IEEE Standard—Adoption of International Standard ISO/IEC 12119: 1994(E)—Information Technology—Software Packages—Quality Requirements and Testing, IEEE Std 1465-1998 (Jun 25), IEEE Press, New York, NY, 1998.

[33] IEEE Recommended Practice for Architectural Description of Software Intensive Systems, IEEE Std 1471-2000 (Sep 21), IEEE Press, New York, NY, 2000.

[34] IEEE Guide—Adoption of PMI Standard—A Guide to the Project Management Body of Knowledge, IEEE Std 1490-2003 (Dec 10), Replaces 1490-1998 (Jun 25), IEEE Press, New York, NY, 2003.

[35] EIA/IEEE Interim Standard for Information Technology—Software Life Cycle Processes—Software Development: Acquirer–Supplier Agreement, IEEE Std 1498-1995 (Sep 21), IEEE Press, New York, NY, 1995.

[36] IEEE Standard for Information Technology—Software Life Cycle Processes—Reuse Processes, IEEE Std 1517-1999 (Jun 26), IEEE Press, New York, NY, 1999.

[37] IEEE Standard for Software Life Cycle Processes—Risk Management, IEEE Std 1540-2001 (Mar 17), IEEE Press, New York, NY, 2001.

[38] IEEE Recommended Practice for Internet Practices—Web Page Engineering—Intranet/Extranet Applications, IEEE Std 2001-2002 (Jan 21, 2003), IEEE Press, New York, NY, 2003.

[39] Industry Implementation of International Standard ISO/IEC 12207:1995 Standard for Information Technology—Software Life Cycle Processes—Software Life Cycle Processes, IEEE/EIA 12207.0/.1/.2-1996 (Mar), IEEE Press, New York, NY, 1996.

[40] IEEE, IEEE Standards Collection, Software Engineering, 1994 Edition, IEEE Press, New York, NY, 1994.

[41] IEEE, IEEE Software Engineering Standards Collection, IEEE Press, New York, NY, 2003.

[42] IEEE, Software and Systems Engineering Standards Committee Charter Statement, http://standards.computer.org/S2ESC/S2ESC_pols/S2ESC_Charter.htm, 2003.

[43] S2ESC Guide for Working Groups, http://standards.computer.org/S2ESC/S2ESC_wgresources/S2ESC-WG-Guide-2003-07-14.doc, 2003.

[44] IEEE, Guide to the Software Engineering Body of Knowledge (SWEBOK), Trial Version, IEEE Press, New York, NY, 2001.

[45] McConnell, S., "The Art, Science, and Engineering of Software Development, *IEEE Software Best Practices,* Vol. 15 No.1, 1998.

ISO PUBLICATIONS

[46] International Standard 9000, Quality Management Systems—Fundamentals and Vocabulary, ISO 9000:2000(E), 2000, Switzerland.

[47] International Standard 9001, Quality Management Systems—Requirements, ISO 9001:2000(E), 2000, Switzerland.

[48] International Standard 90003, Software and System Engineering—Guidelines for the Application of ISO 9001:2000 to Computer Software, ISO/IECF 90003:2003(E), 2003, Switzerland.

[49] International Standard 9004, Quality Management Systems—Guidelines for Performance Improvements, ISO 9004:2000(E), 2000, Switzerland.

[50] ISO 9000 Introduction and Support Package Module: Guidance on ISO 9001:2000 Subclause 1.2 "Application," 524R4; ISO/TC 176/SC 2, 2004, London.

[51] ISO 9000 Introduction and Support Package Module: Guidance on ISO 9001:2000

Guidance on the Documentation Requirements of ISO 9001:2000, 525R, ISO/TC 176/SC 2, 2001, London.

[52] ISO 9000 Introduction and Support Package Module: Guidance on ISO 9001:2000 Guidance on the Terminology Used in ISO 9001:2000 and ISO 9004:2000, 526R, ISO/TC 176/SC 2, 2001.

[53] ISO 9000 Introduction and Support Package Module: Guidance on ISO 9001:2000 Guidance on the Concept and Use of the Process Approach for Management Systems, 544R2, ISO/TC 176/SC 2, London, 2004.

[54] ISO 9000 Introduction and Support Package Module: Guidance on ISO 9001:2000 Guidance on Outsourced Processes, 630R2, ISO/TC 176/SC2, London, 2003.

[55] ISO/IEC 15504 International Standard for Software Process Assessment, ISO/IEC TR 15504, Switzerland, 2003/2005.

[56] ISO 19011 Guidelines for Quality and Environmental Management Systems Monitoring, BS EN ISO 19011:2002, London, 2002.

[57] Information technology—Software Measurement—Functional Size Measurement—Definition of Concepts, ISO/IEC JTC1/SC7, ISO/IEC 14143.1:1999, Canada, 1998.

[58] Systems Engineering—System Life Cycle Processes, ISO/IEC JTC1/SC7, ISO/IEC 15288:2002, Canada, 2002.

[59] Information Technology—Software Engineering—Software Measurement Process, ISO/IEC JTC1/SC7, ISO/IEC 15939:2001, Canada, 2001.

OTHER REFERENCES

[60] Arthur, L., *Software Evolution: The Software Maintenance Challenge,* Wiley, 1988.

[61] Babich, W., *Software Configuration Management,* Addison-Wesley, 1986.

[62] Victor R. Basili, et al., "A Reference Architecture for the Component Factory," *ACM Trans. Software Eng. and Methodology,* Vol 1., No. 1, Jan. 1992, pp. 53–80.

[63] Bersoff, E., Henderson, V., and Siegel, S., *Software Configuration Management: A Tutorial,* IEEE Computer Society Press, 1980, pp. 24–32.

[64] Bounds, N. M. and Dart S. A., *Configuration Management (CM) Plans: The Beginning to Your CM Solution,* Software Engineering Institute, Carnegie Mellon University, 1993.

[65] Bredemeyer Consulting, *The Architecture Discipline—Software Architecting Success Factors and Pitfalls,* http://www.bredemeyer.com/CSFs_pitfalls.htm, 2004.

[66] Carr, M. et al., *Taxonomy-Based Risk Identification,* Software Engineering Institute, Carnegie Mellon University, Technical Report, CMU/SEI-93-TR-006, 1993.

[67] Croll, P., "Eight Steps to Success in CMMI-Compliant Process Engineering, Strategies and Supporting Technology," in *Third Annual CMMI® Technology Conference and Users Group,* 2003.

[68] Croll, P., "How to Use Standards as Best Practice Information Aids for CMMI-Compliant Process Engineering," in *14th Annual DoD Software Technology Conference,* 2002.

[69] Croll, P. and Land, S. K., "S2ESC: Setting Standards for Three Decades," *IEEE Computer Magazine,* January 2005.

[70] Davis, A., *Software Requirements: Analysis and Specification,* Prentice-Hall, 1990.

[71] U.S. Department of Defense, *Software Transition Plan,* Data Item Description DI-IPSC-81429.

[72] U.S. Department of Energy Quality Managers Software Quality Assurance Subcommittee, "Software Risk Management, A Practical Guide," SQAS21.01.00—1999, 2000.

[73] Department of Justice Systems Development Life Cycle Guidance, Interface Control Document Template, Appendix C-17; http://www.usdoj.gov/jmd/irm/lifecycle/table.htm.

[74] Dunn, R. H. and Ullman, R. S., *TQM for Computer Software,* 2nd edition, McGraw-Hill, 1994.

[75] Ford, G. and Gibbs, N., "A Mature Profession of Software Engineering," Software Engineering Institute, Carnegie Mellon University, Technical Report, CMU/SEI-96-TR-004, 1996.

[76] Freedman, D. P. and Weinberg, G. M., *Handbook of Walkthroughs, Inspections, and Technical Reviews, Evaluating Programs, Projects, and Products,* 3rd edition, Dorset House, 1990.

[77] Gremba, J. and Myers, C., "The IDEAL Model: A Practical Guide for Improvement," *Bridge,* Issue 3, Software Engineering Institute, Carnegie Mellon University, 1997.

[78] Hass, A. M. J., *Configuration Management Principles and Practice,* Addison-Wesley, 2003.

[79] Jalote P., *An Integrated Approach to Software Engineering,* 2nd edition, Springer-Verlag, 1997.

[80] Kasunic, M., "An Integrated View of Process and Measurement," Software Engineering Institute, Carnegie Mellon University, Presentation, http://www.sei.cmu.edu/sema/pdf/integrated-view-process.pdf, 2004.

[81] Land, S. K., "1st User's of Software Engineering Standards Survey," IEEE Software and Systems Engineering Standards Committee (S2ESC), 1997.

[82] Land S. K., "2nd User's of Software Engineering Standards Survey," IEEE Software and Systems Engineering Standards Committee (S2ESC), 1999.

[83] Land, S. K., "IEEE Standards User's Survey Results," in *ISESS '97 Conference Proceedings,* IEEE Press, 1997.

[84] Land, S. K., "Second IEEE Standards User's Survey Results," in *ISESS '99 Conference Proceedings,* IEEE Press, 1999.

[85] Land, S. K., *Jumpstart CMM®/CMMI® Software Process Improvement Using IEEE Software Engineering Standards,* John Wiley, 2004.

[86] McConnell, S., *Professional Software Development,* Addison-Wesley, 2004.

[87] McFeeley, B., *IDEAL: A User's Guide to Software Process Improvement,* Software Engineering Institute, Carnegie Mellon University, Handbook, CMU/SEI-96-HB-001, 1996.

[88] Moore, J., "Increasing the Functionality of Metrics through Standardization," in *Conference on Developing Strategic I/T Metrics,* 1998.

[89] Moore, J., *Road Map to Software Engineering—A Standards Based Guide,* Wiley, 2005.

[90] Phifer, E., *DAR Basics: Applying Decision Analysis and Resolution in the Real World,* Software Engineering Institute, Carnegie Mellon University, SEPG Presentation, http://www.sei.cmu.edu/cmmi/presentations/sepg04.presentations/dar.pdf, USA, 2004.

[91] Pressman, R., *Software Engineering,* McGraw-Hill, 1987.

[92] Royce, W., "CMM® vs. CMMI, From Conventional to Modern Software Management," *The Rational Edge,* 2002.

[93] Schach, S. R., *Classical and Object-Oriented Software Engineering,* 3rd Edition, Irwin, 1993.

[94] SEI, *The Capability Maturity Model: Guidelines for Improving the Software Process,* v.1.1, Software Engineering Institute, Carnegie Mellon University, 1997.

[95] SEI, *A Framework for Software Product Line Practice, V 4.2,* Software Engineering Institute, Carnegie Mellon University, Web Report, http://www.sei.cmu.edu/plp/framework.html#outline, 2004.

[96] SEI, *Capability Maturity Model Integration (CMMI) for Software Engineering,* v.1.1 Staged Representation, Software Engineering Institute, Carnegie Mellon University, Technical Report, CMU/SEI-2002-TR-029, 2002.

[97] SEI, *Integrated Product Development (IPD)-CMM®,* Software Engineering Institute, Carnegie Mellon University, Model Draft, 1997.

[98] SEI, *Organizational Process Improvement Checklist Using IDEAL^{SM},* Software Engineering Insititute, Carnegie Mellon University, Presentation, www.sei.cmu.edu/ideal/ideal.present.

[99] USAF Software Technology Support Center (STSC), *CMM-SE/SW V1.1 to SW-CMM® V1.1 Mapping,* 2002.

[100] Veenendall, E. V., Ammerlaan, R., Hendriks, R., van Gensewinkel, V., Swinkels, R., and van der Zwan, M., "Dutch Encouragement; Test Standards we Use in Our Projects," *Professional Tester,* Number 16, October 2003.

[101] Westfall, L. L., "Seven Steps to Designing a Software Metric," BenchmarkQA, Whitepaper, 2002.

[102] Whitgift, D., *Methods and Tools for Software Configuration Management,* Wiley, 1991.

[103] Wiegers, K. E., *Creating A Software Engineering Culture,* Dorset House Publishing, 1996.

[104] Williams and Wegerson, *Evolving the SEPG to a CMMI World,* http://www.sei.cmu.edu/cmmi/presentations/sepg03.presentations/williams-wegerson.pdf, 2002.

[105] Zubrow, D., Hayes, W., Siegel, J., and Goldenson, D., *Maturity Questionnaire,* Special Report, CMU/SEI-94-SR-7, Software Engineering Institute, Carnegie Mellon University, 1994.

Index

About the Authors

Susan K. Land

Ms. Land is a Program Manager with the Intelligence Division of NGIT/TASC, Huntsville Operations. Ms. Land has over 19 years of Information Technology work experience in the areas of information management systems programming, database systems development, and enterprise web-based application programming. She is an acknowledged expert in the field of software engineering standardization, software process improvement, and software engineering management. Ms. Land is an IEEE Computer Society (CS) Certified Software Development Professional (CSDP) certificate holder and holds an additional 13 certifications on a variety of topics supporting specific software development methodologies. Ms. Land is a TASC President's Coin recipient and a Northrop Grumman Technical Fellow.

Ms. Land is the Production Planner for the $170M America's Army (AA) Program (www.americasarmy.com). America's Army is a widely distributed online game produced by the U.S. Army. This program is supported by the America's Army Project Office at the Software Engineering Directorate (SED), Redstone Arsenal, AL. This Project Office is responsible for providing software engineering management and oversight for all AA development.

Ms. Land, a Senior Member of IEEE, has contributed her personal time in support of the development and definition of software engineering as a profession. She has served in a number of IEEE Computer Society (CS) offices including 2nd Vice President, as a member of the Board of Governors (BoG), and as Vice President for IEEE CS Standards Activities. She is also Chair of the Technical Activities Awards Committee, a member of the Computer Society International Design Competition (CSIDC) Committee, Chair of the Software Engineering Portfolio Oversight Committee (SEPOC), member of the Information Technology Oversight Committee (ITOC), and past participant of many other IEEE CS Committees and Boards. In 2006 she was honored to be included among the IEEE CS Golden Core and as a member of the Distinguished Visitors Program.

John W. Walz

Mr. Walz retired as Senior Manager, Supply Chain Management, Lucent Technologies. His 30 year career at Lucent and AT&T was highlighted by customer-focused and dynamic results-oriented management with an excellent technical background and over 20 years of management/coaching experience. John held leadership positions in hardware and software development, engineering, quality planning, quality auditing, quality standards implementation, and strategic planning. He was also responsible for Lucent's Supply Chain Network quality and a valuable team member for Lucent's quality strategic planning. This work resulted in many locations achieving ISO 9001 and then TL 9000 registrations. These organizations achieved and shared quality and process improvement results at annual Lucent sharing rallies and external best practices conferences. Mr. Walz actively participated in both software engineering and quality system standards development and implementation. Mr. Walz also wrote the TL 9000 Chapter in the *ISO 9000 Handbook,* 4th edition. Mr. Walz is an expert speaker in the Distinguished Visitor Program of the IEEE Computer Society. John earned his BSEE and MSEE degrees from Ohio State University.

Additional titles available by these authors:

Jumpstart CMM®/CMMI® Software Process Improvement, Using IEEE Software Engineering Standards
Practical Support for CMMI-SW Software Project Documentation, Using IEEE Software Engineering Standards